D0710395

Maintenance Planning, Scheduling,

and

Coordination

Don Nyman
and
Joel Levitt

Industrial Press Inc.

New York

Library of Congress Cataloging-in-publication Data

Nyman, Don.
  Maintenance planning, scheduling, and coordination/ Don Nyman, Joel Levitt.
  p.cm
  ISBN 0-8311-3143-8
  1. Product life cycle. 2. Production management. I. Levitt, Joel, 1952-II. Title.

TS176 .N96 2001
658.5--dc21

2001024460

Industrial Press Inc.
989 Avenue of the Americas
New York, NY 10018

First Edition, August 2001

Sponsoring Editor: John Carleo
Interior Text and Cover Design: Janet Romano

Printed in the United States of America

10 9 8 7 6

Dedication

This book is dedicated to our wives (Barbara Spender Nyman and Barbara Cavanaugh Levitt) and our families for all their love, patience, understanding, support, and sacrifices during the several decades we have pursued our careers; often away from home a majority of the weeks in each calendar year.

This book is also dedicated to you maintenance professionals who labor to keep our factories running, our airports open and our trucks on the road. You do your magic without visibility, without a lot of status, and sometimes without even a thank you. We would like to acknowledge your dedication, your hours, and your sense of duty.

Don Nyman
Joel Levitt
August 2001

OTHER TITLES BY JOEL LEVITT

INTERNET GUIDE FOR MAINTENANCE MANAGEMENT

1998, 160 pages, illus., ISBN 0-8311-3081-4

Similar in purpose to other successful Internet guides, this is a unique guide to using the Internet for maintenance managers and all managers concerned about or affected by the maintenance of assets. The material is organized in 11 brief, accessible chapters that have been written in a style and indexed in a way that makes it easy for readers to quickly locate the information they need.

THE HANDBOOK OF MAINTENANCE MANAGEMENT

1997, 476 pp., ISBN 0-8311-3075-X

This unusually comprehensive book is intended to be used in different ways by different groups. It was designed as a complete survey of the field for students or maintenance professionals, as an introduction to maintenance for non-maintenance people, as a review of the most advanced thinking in maintenance management, as a manual for cost reduction, as a primer for the stockroom, and as an element of a training regime for new supervisors, managers and planners. The author accordingly presents a customized curriculum--or road map--for many of these groups of readers.

MANAGING FACTORY MAINTENANCE

1996, 290 pp., ISBN 0-8311-3063-6

Tap into Joel Levitt's vast array of experience and learn how to improve almost any aspect of your maintenance organization (including your own abilities). Beginning with the twenty steps that are necessary to achieve world class maintenance, the book then investigates how to compare your organization to industry standards. Skills required for effective maintenance management are then studied including such unusual but critical topics as managing work requests, work orders, and repair history documents. Both preventive and predictive maintenance are examined in depth. The chapter on predictive inspection provides examples of every major technology including oil and vibration analysis and temperature and ultrasonic inspection. How to improve your system by implementing TPM, CMMS, RCM, and Maintenance Quality Improvement is full yexplained and demonstrated. Topics unique to this booke are a self-evaluation clinic that allows you to compare your personality with over 500 of your peers and a section on time management that will improve your management skills.

Preface

One intention of this book is to aid organizations that pursue "Maintenance Excellence." Maintenance excellence is that state of maintenance management and performance that effectively applies the leading edge policies, procedures, systems, structures, methods, and technologies to maintenance. Maintenance Excellence is requisite to the achievement of World-Class Operations (an organization that is competitive with the best in the world).

Well-planned, properly scheduled, and effectively communicated jobs accomplish more work, more efficiently, and at a lower cost. Work properly prepared in this fashion disturbs operations less frequently, and is accomplished with higher quality, greater job satisfaction, and higher organizational morale than jobs performed without proper preparation.

Without proper scheduling, the crucial proactive routines optimized through other vital techniques (RCM, Predictive Maintenance, and Condition Based Maintenance) most likely will not be performed when due.

The key to achieving maintenance excellence is nothing new. It has always been and still remains: get the basics right and make maintenance excellence a goal of the entire organization. Foremost among the basics commonly contained within world-class programs are planning, parts acquisition, work measurement, coordination and scheduling. Together these basics constitute the preparation required for effective execution of maintenance work. Throughout this book we will use the term *"Job Preparation"* when speaking of these distinct tasks in their totality.

Our book focuses on and deals specifically with these preparatory tasks that lead to effective utilization and application of maintenance resources. Other aspects addressed include maintenance management, performance and control are discussed as necessary to clarify scope, responsibilities and contributions of the Planner/Scheduler function, and other functions impacted by or supportive of Job Preparation, Execution, and Completion.

The larger the maintenance organization, the more likely there are planners to be trained. But, regardless of size, every maintenance organization needs to plan for the effective execution of its workload. Thus, the book is a vital training document for planners as well as an educational document for those to whom planners are responsible. The book is also important for those who interface with the planning and scheduling function and are dependent upon the many contributions of planning and scheduling for operational excellence.

Because we are chronicling maintenance management basics, this book is not an original work, but a compilation of knowledge that is already in the public domain. We are therefore indebted to many associates, clients, and others who have worked the field of maintenance throughout the decades.

This book has been designed to be used in two ways. Primarily it has been written to be read by maintenance professionals and others from the beginning to the end. Both authors teach these topics extensively, so the book is a course in planning and is to be used as such. It was written to tell the whole story of maintenance planning from the beginning to the end.

For experienced practitioners of planning, the book can be used as a reference tool to clarify one of the planning steps or to provide ideas about forms or control structures.

Table of Contents

Introduction

Because maintenance is an increasing portion of operational costs, it is seen as the last opportunity for major operational improvement. Most organizations continually search for any means (program, process, concept, or approach) by which to improve their maintenance function. They strive to assure that each maintenance dollar is well spent (labor productivity) while achieving equipment reliability (asset productivity). Organizations must be successful in this quest, if they are to survive and thrive.

Maintenance organizations everywhere have the responsibility to assure optimum use of the capacity of the enterprise. The entire organization becomes maintenance conscious whenever the maintenance job is not properly or efficiently performed; unsafe operations, production downtime, quality problems and the loss of heat, light, and other utilities do not go unnoticed.

It may help if reactive maintenance is visualized as falling into a deep, dark, dank, slimy pit of deferred maintenance. The fall doesn't end before hitting rock bottom, which equates to bankruptcy, outsourcing, or possibly privatization. Before positive progress can be made, one must first climb out of the pit. Preventive/Predictive Maintenance (PPM) is not conceived to put equipment in proper condition, but to maintain it in that condition from the time of acquisition or restoration.

Proactive maintenance requires a cultural transition from a reactive to a proactive environment. Successful change demands an *"integrated maintenance management partnership"*. This concept is illustrated in the **Maintenance Arch** shown in Figure 1. Of the many maintenance activities and functions here depicted, planning, coordination, and scheduling make the most profound contribution to this cultural change and have the greatest impact on timely and effective accomplishment of maintenance work.

ix

Introduction

Planning must be at the core of the maintenance effort because it provides for reliable delivery of all the other proactive programs. Combined with PM/PdM engineered through RCM, these programs produce quantum benefits that accrue to the enterprise's bottom line.

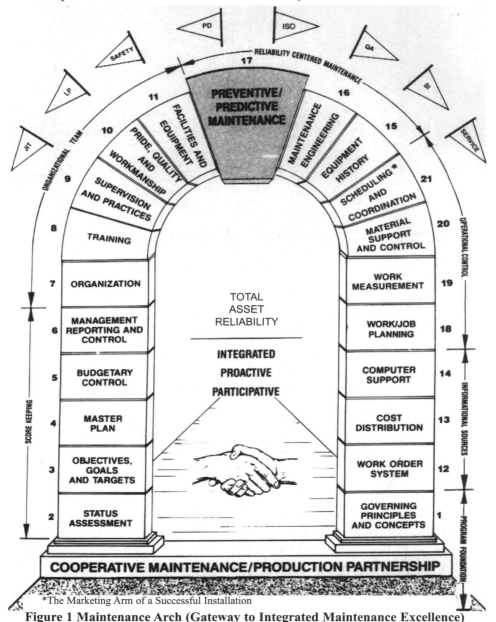

*The Marketing Arm of a Successful Installation

Figure 1 Maintenance Arch (Gateway to Integrated Maintenance Excellence)

Introduction

THE MAINTENANCE ARCH

The components that make up the maintenance arch represent the integration of twenty-one building blocks that are essential to the achievement of maintenance excellence in support of world-class operations through asset/capacity reliability. The concept is built upon a bedrock of sustained management commitment, support, and involvement, and it is laid upon a footer of an operations/maintenance partnership. Maintenance cannot achieve the objective alone. Realization requires the commitment of the entire organization.

Maintenance is a resource that contributes to the achievement of profit objectives. The several banners that fly from the top of the arch convey this concept. These acronyms and words represent some of the many management initiatives that are maintenance dependent; Just In Time, Loss Prevention, Employee Safety, Avoidance of Property Damage, Certification, Quality Assurance, Avoidance of Business Interruption, Customer Service, among others.

The building blocks in the arch are interdependent, which means that they are linked. It is therefore difficult to undertake improvement of one without the addressing needs associated with the others.

◆ As in many arenas, successful maintenance processes begin with shared beliefs (Governing Principles and Concepts). Status Assessment measures the current state relative to those beliefs. The Master Plan sets forth the actions, responsibilities, and time lines, and essential resources to close the gap between the current state and established Objectives, Goals, and Targets. Budgetary refinement is required to support the Master Plan. Management Reporting is the feedback vehicle by which sustained management commitment is earned.

◆ Preventive/Predictive Maintenance is the keystone of the Arch and is the vehicle by which reliability is assured. It requires a strong Maintenance Engineering function using Equipment History and Reliability Centered Maintenance to optimize the PM/PdM process.

◆ The Organization must be structured for proactive rather than reactive response. Skills Training and Facilities, Tools and

Introduction

Equipment in which the organization can take Pride are essential if Supervision is to achieve functional Quality Assurance and adherence to policies and procedures.

◆ The preparatory trio achieves effective utilization of maintenance resources. Planning, Coordination, and Scheduling supported by a strong Computerized Work Order System with reliable Cost Distribution and Work Measurement are needed to fulfill challenging expectations.

DEFINITIONS

There is often confusion between the functions of planning, coordination and scheduling. These three parts (planning, coordination, and scheduling) are closely related and are usually performed by the same individual(s), but they are distinct activities. Part of the problem is that these three functions are compressed together, particularly on smaller jobs.

Planning (how to do the job – Chapters 7 through 11): Planning is the development of a detailed program to achieve an end (i.e., a maintenance repair or rebuild). It is the advanced preparation of a specific job so it can be performed in an efficient and effective manner.

Figure 2 The Planner's Role

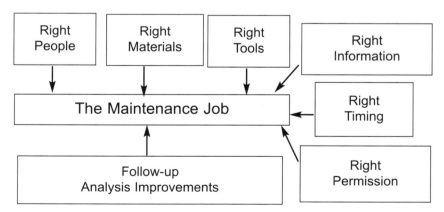

Introduction

Planning insures all that necessary logistics have been coordinated for the job execution phase to take place at a future date. Planning is a process of detailed analysis that determines and describes the work to be performed, the sequence of associated tasks, methods to be used for their performance, and the required resources – including skills, crew size, man-hours, parts, materials, special tools, and equipment, and an estimate of total cost. It also includes identification of safety precautions, required permits, communication requirements, and reference documents such as drawings and wiring diagrams. It addresses essential preparation, execution and start-up efforts. Work Measurement (the setting of job duration and labor estimates) and activation of required procurements, are part of the planning process.

Coordination (Chapter 12): Encompasses the logistical efforts of assembling all necessary resources so the job is ready to be scheduled. It is accomplished in harmony with Purchasing, Receiving, and Stores. Operations, Engineering, and Maintenance then review all jobs ready to be executed and agree upon which are the most important jobs to be performed during the coming week – given the limitation of available resources and feasibility of releasing involved assets to Maintenance.

Scheduling (when to do the job – Chapter 13): Scheduling is the written process whereby labor resources and support equipment are allocated/appointed to specific jobs at a fixed time when Operations can make the associated equipment or job site accessible.

As part of job execution (Chapter 14), the responsible maintenance supervisor assigns the specific job to the appointed individual.

The resulting schedule represents what the organization expects to accomplish with the maintenance resources that will be available and paid for during the coming week. "We all accomplish more when working to achieve a published expectation." Scheduling is the marketing arm of a successful maintenance management installation because it yields the earliest benefits visible to internal customers.

Considered together, these three activities constitute *"Job Preparation."* They are supportive functions distinct from line supervision (which is responsible for oversight of job execution) and are best performed by para-professional, management personnel. They are predicated on the principle that; "Maintenance management will achieve its best results when each mechanic is given specific tasks to be completed in a definite time period (scheduling) in a specific manner (planning)." Mechanics do not plan for their own efficiency. *"The productivity of work is not the responsibility of the worker but of the manager"* (Peter Drucker). The supervisor, the worker, and their colleagues should each know what is expected – including the goal and target for completion of each job.

WHY PLAN, COORDINATE AND SCHEDULE MAINTENANCE JOBS?

Most maintenance departments do not plan to fail; they simply fail to plan and therefore do indeed fail. The major reason behind failure to plan is that putting out today's fires is given priority over planning for tomorrow – thereby ensuring that future equipment failures will require reactive response (ample supply of kindling to be consumed in future reactive fires). Secondary things are constantly put first by default. This may not be bad—but when we *get down to putting the twelfth job first, we are in trouble*. Reactive maintenance is simply a vicious circle, a continuous downward spiral.

We plan because planned maintenance reduces waiting and delay times that mechanics inevitably encounter when performing work that has not been properly prepared for. Poor utilization of a mechanic's time is usually not his or her fault. Nothing is more detrimental to maintenance performance and morale than unplanned or poorly planned jobs. All enthusiasm is lost when a crew arrives at a job site to find that the work can neither be started nor completed due to lack of proper preparation. Mismanagement is highly visible to the workers on the front lines. Conversely, with effective preparation, maintenance crews can take pride in their work and organizational performance is improved.

Most of us think we work best under pressure. Unconsciously we procrastinate until pressure forces us into reaction. But, do we really work best under pressure, or do we simply work faster and less effectively due to lack of preparation and hasty judgments?

Although urgency can engulf managers, urgent tasks are not always the most important tasks to be performed. The tyranny of urgency lies in distortion of priorities (subtle cloaking of minor projects with major status under the guise of crisis). But, the measure of true management is the ability to distinguish important from urgent, refusal to be tyrannized by the urgent, refusal to manage by crisis. Time must not be simply consumed but must be well utilized.

Introduction

Granted, urgent tasks (repair of breakdowns) call for instant action, whereas important tasks (PM/PdM's and planned backlog relief) rarely must be done today, or even this week. The momentary appeal of urgent tasks and instant action is irresistible and so they devour our resources and energy. But in time, their deceptive prominence fades. With a sense of loss, we recall the important tasks that were pushed aside and realize that we are slaves to urgency. A major obstacle to organizational success is allowing urgency alone to determine how resources will be consumed ... without considering the importance of the activity relative to the broad objectives of the enterprise. To climb out or the "reactive pit" adequate resources must remain to perform the important work of reliable routine services (PM/PdM) and timely backlog relief. For they make us better in the future, thereby reducing future failures requiring emergency response.

Advanced planning, coordination, scheduling, and the pursuit thereof are proactive skills. Planners must be allowed to focus on "tomorrow" rather than being caught up in "today." Well-planned, properly scheduled and effectively coordinated jobs can be accomplished more efficiently, at lower cost, with fewer disturbances of operations, higher quality (reduced variability of processes), improved morale with greater job satisfaction, increased longevity of equipment, and reduced parts usage. More work is completed more promptly with fewer resources, thereby increasing customer service and profit. Studies show that 3-5 hours of execution time can be saved for every hour of advanced preparation. Planned jobs require only half as much time to execute, as that needed for unplanned jobs.

When long-range objectives become obscured it is easy to replace them with much shorter-range and even hopelessly misplaced goals, such as speed of repair. This argument does not decry being efficient in the right things at the right time, but speed, as an end in itself, is futile.

Therefore, avoid overstressing efficiency while neglecting the careful selection of areas in which efficiency is to be sought.

Efficiency's myth lies in the assumption that "fast" is, ipso facto, effective; and the harder one works, the more he accomplishes. The allure of speed seems irresistible; who could be against it? Rapid performance of the wrong task is not effective at all. Therefore, the "hazard" of emphasizing speed without regard to effectiveness must be avoided at all costs.

Action that puts speed ahead of results but disregards organizational objectives, in the long run, will be totally ineffective and likely will need to be done over. Doing the job right the first time requires proper preparation. If you don't have time to do it right, when will you have time to do it over?

LEARN TO SAY "NO"

Every job or project has an advocate of priority somewhere. Because people do not like to say no, everybody's urgency takes priority and nothing of importance is accomplished. The problem is not usually with priorities one through three, but with "posteriorities" four through infinity. One simply cannot achieve excellence of performance without concentrating effort in the critical areas.

One has to say about a suggested, but secondary project "this should be done, but it's not our first priority. If it must be done we'll have to let somebody else do it. Do not be timid about using the greatest time-saving word in the English language, that little two-letter word "No." Decide what not to do. Bill Cosby has been heard to say "I don't know the key to success but the key to failure is trying to please everybody."

Keeping overall objectives in the forefront allows us to schedule time based on long-term objectives rather than on immediate crises.

In contrast to the benefits of proactive maintenance, the following sonnet effectively conveys the curse associated with reactive maintenance.

Figure 3 The Rush Job

I am a rush job.
I belong to no age, for man has always hurried.
I prod all human endeavors.
Men believe me necessary—but falsely.
I rush today because I was not planned yesterday.
I demand excessive energy and concentration.
I override obstacles, but at great expense.
I illustrate the old saying "Haste makes waste."
My path is strewn with the evils of overtime,
mistakes, and disappointment.
Accuracy and quality give way to speed.
Ruthlessly I rush on ...I am a rush job.

Anonymous

Introduction

OBJECTIVES OF WORK PREPARATION

Work Preparation has several primary objectives. It:

- ❏ Forms the communication center from which all maintenance activity is communicated and coordinated.
- ❏ Optimally supports the operations by improving maintenance in the broadest sense, considering both the technical aspects and the service provided to the internal customer. Do not neglect customer relations. Loss of production capacity is minimized, and maintenance work is completed, when needed, in a safe, environmentally conscious environment, at optimal cost.
- ❏ Is requisite to effective performance of the maintenance function.
- ❏ Establishes equitable allocation of resources based on varying business needs, demands of customer service (internal as well as external), and other criteria.
- ❏ Coordinates use of the various input resources (material, manpower, and equipment), as required, to complete each job in an orderly manner and at least overall cost.
- ❏ Achieves optimal, effective, utilization and application of all maintenance resources with minimal manpower delay and idleness.
- ❏ Reflects the foresight to anticipate and thereby forestall delays that commonly plague maintenance work performed reactively. Labor, materials, and equipment needs are identified on paper and provided for before resources are committed to the field.
- ❏ Allows a maintenance craftsman to prepare for, safely perform, and economically complete, each job to the satisfaction of the requestor with minimal interruption of operations.
- ❏ Establishes expectancy against which utilization of maintenance resources can be evaluated. Individuals know what is expected of them and output and productivity of maintenance workers is favorably impacted.
- ❏ Relieves supervisors of much indirect activity, enabling them to spend more time at the job site leading job execution efforts.

Without proper planning and scheduling, maintenance is haphazard, costly, and ineffective, and maintenance will consistently fail to meet promised dates. These failures will cause constant problems for Operations. When Maintenance consistently fails to put equipment back into service as promised, Operations will become increasingly reluctant to release equipment in the future.

Introduction

PREREQUISITES

Realization of maintenance objectives requires adherence to proven prerequisites, principles, and procedures. The following must be in place:

❑ Lead-time is essential. Without it, there is no time for the planning process. Needed work must be identified as far in advance as possible so that the backlog of work is known and jobs can be effectively planned prior to scheduling their execution.

❑ A strong Work Order System must be instituted because control of maintenance work is instituted through the work order system.

❑ An Organizational Structure that fosters Proaction rather than Reaction is needed with provision within the Operating Plan for proaction (Chapter 3).

❑ Reasonable spans of control and proper job scopes for supervisors, planners, and maintenance/reliability engineers. In turn, these positions must be staffed by qualified personnel … not castoffs.

❑ Understanding of the department's mission in relation to company objectives and certainty that each project undertaken fits within and contributes to the fulfillment of "The Mission."Each job request must be evaluated as to its importance in respect to the operation as a whole. Keeping overall objectives in the forefront allows resources to be scheduled based on long-term objectives rather than immediate crisis. When long-range objectives become obscured, short-range and even hopelessly mis placed goals easily replace them.

❑ Assistance for Operations in establishing a practical level of maintenance whereby long as well as short-term operating plans are met while maintenance needs are accomplished through anticipation, planning, and scheduling. The facilities must then be maintained at specified levels of operating efficiency, at the lowest possible cost consistent with the goals of producing quality product as economically as possible.

❑ Regard for operations as an internal customer, with no neglect of customer service, within the framework of the Production/Maintenance/Quality Partnership. The partnership emphasis forestalls excessive use of "the customer is always right" argument.

Maintenance customers deserve to have their work performed on a timely basis. Therefore, backlogs must be kept within a reasonable limit. Backlogs below minimum do not provide sufficient volume of work to accommodate smooth scheduling. Backlogs above maximum turn so slowly that it is impossible to meet customer needs on a timely basis.

Introduction

Special or heavy demands cannot be met unless excessive backlog is addressed by providing additional resources or by relaxing priorities. Emergency work is performed at the expense of scheduled jobs. The scheduled jobs displaced result in work being carried over to the next schedule period unless addressed by a temporary increase in resources, such as overtime.

❏ Effective planning, coordination, and scheduling of maintenance work must be completed far enough in advance to permit Operations to plan for equipment to be out -of-service so as to minimize nonproductive time and pro duction shortfalls.
Prior to execution, all shutdown work must be reviewed with key production personnel so that their intimate and expert knowledge can be fully utilized. Based on authorized requests, required work must be defined and executed on time and with quality workmanship (knowing what is to be done, when, and how best to do it   and then doing it right the first time). Regular feed back regarding status of work requests and completion promises must be provided.

❏ Effective management of materials (parts and supplies) is essential to support efficient and effective job execution.

❏ Supervisors must be in attendance at job sites to provide effective leadership during execution, follow-up, and feed back, assuring that crews make full use of preparatory efforts and perform proper work correctly.

❏ Constant efforts are needed to improve maintenance work methods, completeness, neatness, quality, and efficiency.

❏ Engineering support of maintenance efforts is vital to address the root causes of repetitive failures.

❏ Accountability for the level of cost incurred in the performance of requested maintenance is essential.

Each hour of effective planning typically returns three to five hours in mechanic time or equivalent savings (measured in cost of material and production downtime). With more work done more promptly and with greater effectiveness, customer service and asset reliability improve dramatically.

Continuation of a reactive culture would require worldwide adoption of a new weekly calendar as depicted in figure 4.

Introduction

Figure 4 The Rush Job Calendar

NEG	FRI	FRI	THU	WED	TUE	MON
8	7	6	5	4	3	2
16	15	14	13	12	11	9
23	22	21	20	19	18	17
31	30	29	28	27	26	24
38	37	36	35	34	33	32

1. Every job is in a rush. Everyone wants his or her job yesterday. With this calendar, a customer can order the job on the 7th and have it delivered on the 3rd.
2. All customers want their jobs on Fridays so there are two Fridays in each week.
3. There are seven extra days at the end of the month for those end-of-the-month jobs.
4. There will be no first-of-the-month bills to be paid, as there isn't any first. The tenth and the twenty-fifth have been omitted—in case you have been asked to pay on one of those days.
5. There are no bothersome and non-productive Saturdays and Sundays. No time-and-a-half or double-time to pay.
6. There is a new day each week called Negotiation Day.

Selling Planning, Coordination, and Scheduling to Management and Operations

How can the crucial maintenance functions of planning, coordination and scheduling be sold to management and how can all departments be convinced to follow the procedures necessary to capture the full benefit of planning and scheduling? The first challenge is to gain managerial approval to fund planner positions, train the staff, and build databases of crucial support information. The second challenge is to gain true commitment from Operations, Purchasing, Storeroom, and other organizational units.

SELLING MANAGEMENT

Within most organizations, the battle for funding is extremely competitive. Maintenance begins with two strikes because the function is poorly understood and therefore under appreciated. Why invest in a function that is viewed only as a "necessary evil"? The second strike is that, on the surface, maintenance does not contribute to the "bottom line." In the traditional "Who adds value?" analysis maintenance doesn't seem to add value.

In fact, investments in maintenance yield significant returns and do add value. We must sell our contributions; which are capacity assurance, reliability, and customer satisfaction at lower unit cost. Often we must educate before we can sell. Understanding precedes appreciation. Facts and quantification are necessary to gain managerial attention. This drive for facts justifies the need for the CMMS (Chapter 17).

WORK SAMPLING

When the necessary information to make your case is not available from CMMIS (computerized maintenance management information system), Activity Sampling becomes the alternative vehicle. For the uninitiated, this technique uses random observations of the maintenance work force with categorization by nature of each observation (Figure 1.1). Sufficient observations must be made to gain statistical reliability (approximately 700 per population). This means that if you consider the entire maintenance organization as a single population, only 700 observations are necessary. However, if the desire were to separate mechanics from electricians on each of three shifts, 4200 observations would be required to gain statistical reliability.

When this technique is applied, both the time of day that observation tours are made and the path of the tour must be randomized to avoid bias. If observers travel only the aisles, they will observe a disproportional amount of travel.

Sampling results can be used to show management the amount of technician capacity lost for lack of proper preparation of maintenance jobs. If maintenance is still in the reactive mode, results will approximate those shown in Figure 1.1, which compares typical sampling results for reactive maintenance with results for pro-active environments. The percentages presented reflect a number of studies in a variety of industries.

If the management team will not accept the data presented as being representative of the local environment, a site-specific sampling is needed.

The 35% direct work in reactive mode versus the 65% direct work in pro-active mode provides a simple but clear justification for establishment of a Planning, Coordination, and Scheduling function (Figure 1.2). Two supervisors, each with a ten person crew but without planner support will have only seven full-time equivalents "pulling the wrenches" to complete work.

The same two supervisors with planner support (even if it was established without a headcount increase) will have 12.4 full-time equiva-

2

Figure 1.1 Typical Maintenance Worker's Day - Reactive versus Pro-Active

	Reactive without planning & scheduling	Proactive with planning and scheduling
Receiving Instructions	5%	3%
Obtaining tools and materials	12%	5%
Travel to and from job (both with and without tools and materials)	15%	10%
Coordination delays	8%	3%
Idle at job site	5%	2%
Late starts and early quits	5%	1%
Authorized breaks and relief	10%	10%
Excess personal time (extra breaks, phone calls, smoke breaks, slow return from lunch and breaks, etc)	5%	1%
Subtotal	65%	35%
Direct actual work accomplished (as a percentage of the whole day)	35%	65%

Figure 1.2 Simple Justifications For Planner Position

Two Crews without planning		Two Crews with planning	
2	Supervisors	2	Supervisors
0	Planner	1	Planner
20	Total Craftspeople both crews	19	Total Craftspeople both crews
35%	Direct work percentage	65%	Direct work percentage
7.0	Equivalent Full Time workers	12.4	Equivalent Full Time workers

Improved Output from Planning and Scheduling 77%

lents pulling wrenches. This is a net productivity improvement of 77%, which is a fine return from any investment.

Another way to appreciate the advantage of job planning is to depict what happens within an individual job without planning. Technicians jump into the work without forethought. Shortly they encountered a delay for lack of a spare part, tool, or authorization. This sequence may be repeated several times before the job is completed. In the planned mode, the needs are anticipated and provided for before a technician is assigned. The comparison is graphically presented on the next page.

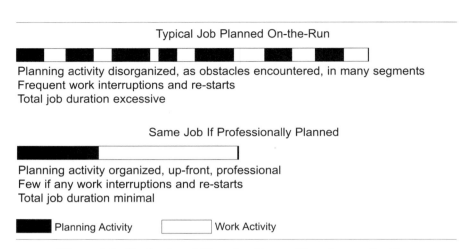

Typical Job Planned On-the-Run

Planning activity disorganized, as obstacles encountered, in many segments
Frequent work interruptions and re-starts
Total job duration excessive

Same Job If Professionally Planned

Planning activity organized, up-front, professional
Few if any work interruptions and re-starts
Total job duration minimal

Planning Activity   Work Activity

Figure 1.3   Professional Planning Versus Planning on the Run

Each dollar invested in planning typically saves three to five dollars during work execution and the duration of a planned job is commonly only half as long as that of an unplanned job. This improvement should be reinvested into "Maintenance Excellence and Asset Reliability," instead of taking it out in the form of cost reduction. This counsel relates to the "Maintenance Iceberg (Figure 1.4)." There are far greater bottom-line contributions to be gained from asset reliability than from "mere" maintenance cost reduction. This approach is also more saleable to the maintenance work force. By committing themselves to asset reliability, they protect the future of their jobs. If the managerial focus is on cost reduction, improved productivity works the maintenance crew out of overtime and possibly their job. This is the common fear of "Labor".

Another issue needs to be clarified. It is necessary to communicate with and sell management in terms they relate to. It is inconsistent with the "Integrated Maintenance Management Partnership" introduced in the Preface of this book to justify each element of the maintenance improvement process individually. The justification for any of the twenty-one building blocks of the "Maintenance Arch" individually is in fact ... **Zero**. Individually, these blocks are just tools. Together they provide a process for asset reliability and continuous improvement; their justification is ample to meet any ROI (return on investment) hurdle.

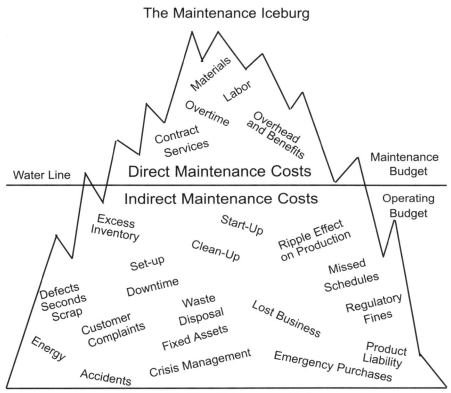

Figure 1.4 The Maintenance Iceberg

SYMPTOMS OF INEFFECTIVE JOB PLANNING

The following frequent symptoms clearly show the ineffectiveness of maintenance resources when operating in a reactive rather than pro-active mode. Delays encountered by mechanics are the norm. For lack of proper prepration, much time is lost:

➢ Gaining detailed knowledge of the required work
➢ Obtaining permits
➢ Identifying and obtaining material, blueprints, tools, and skills required for completion of assigned work
➢ Delivering the above inputs to the job site
➢ Waiting for required spare parts that are not in stock

5

Chapter 1

Due to insufficient thought when requests are written, operating personnel often must alter jobs in progress, causing delays and drop in productivity. Technicians may arrive at a job site on time only to wait for their supervisor, or someone else, to contact the requestor or other operating person for clarification of the requested work. Often the operations department does not have the equipment ready or refuses to release it, despite having previously agreed to the schedule.

The number of craftsmen does not match the magnitude of the job (possibly too many ... possibly too few). Participation of support craft is not anticipated, or workmen are of the wrong craft skill, or have arrived in the wrong sequence (e.g., millwrights arrive before electricians have disconnected equipment).

Craftsmen have no prior knowledge of detailed job task breakdown and are left to decide how to do the job and what materials are required; they then either leave the job site to get materials from the storeroom or stand by and wait for them to be delivered. Even more disruptive is after a job is started, it is discovered that some required parts or materials are not in stock and must be purchased outside the plant (with extra cost for express delivery). The job is halted in a state of disassembly waiting for the items to arrive.

Craftsmen cannot develop work rhythms because of interruptions and delays encountered, causing them to hop from crisis to crisis without completing jobs already in progress.

Maintenance works in standby mode waiting for the next emergency while supervisors become dispatchers of response to breakdowns. Communication is faulty and incomplete throughout the process; taking many guises including failure to satisfy the originator (what was wanted was not what was done), frayed tempers, frustration over wasted efforts, downtime delays, makeshift repairs and prolonged quality problems. All of which result in higher overall costs.

Maintenance personnel in the trenches know the frequency of these occurrences, but managers are often unaware. Mismanagement is highly visible on a daily basis. Shouldn't this be a wake-up call for management?

Chapter 1

CONVEY THE MANY BENEFITS THAT ACCRUE TO EACH STAKEHOLDER

Stakeholders are people in positions having an interest in the way maintenance is conducted. To sell PLANNING, COORDINATION, AND SCHEDULING you must convey the benefits that accrue directly to each stakeholder of the "partnership."

Planning, coordination, and scheduling provide significant benefits to management by:

◆ Providing a central source of information concerning the condition of equipment, the associated maintenance workload, and available resources to perform it.
◆ Improving employee safety
◆ Improving regulatory compliance
◆ Working to achieve the optimal economic level of maintenance in support of both short and long range operational needs
◆ Challenging the need for work requests of questionable value and justification
◆ Accurately forecasting labor and material needs, thus permitting immediate recognition of labor shortages and excesses with steps taken to level peak workloads
◆ Establishing expectations for what is to be accomplished each week by the maintenance payroll investment (forty hours of productive work from each maintenance worker) and analysis of variations from expectations
◆ Improving efficiency through anticipation of needs and avoidance of potential delays
◆ Providing factual data, measurement, analysis of efficiency, and identification of variations from expected performance
◆ Providing information identifying problem areas that require focused attention
◆ Reducing total unit cost of maintenance while improving customer service, condition of equipment and facilities through better use of labor and materials,
◆ Increasing useful life of fixed assets

7

♦ Improving preparation, management, and control of major shutdowns, outages, and turn arounds

Planning, coordination, and scheduling confer significant benefits on operations and production when they:

♦ Provide orderly procedures for requesting, preparing, executing, and closing out maintenance support

♦ Provide close and continual coordination between operations and maintenance with a single point of contact for all emergency and scheduled work whether pending, in-process, or completed

♦ Maintain accurate backlog status

♦ Facilitate anticipation of required repairs before they become emergencies

♦ Coordinate the cooperation necessary to provide essential maintenance resources at times when operations can best relinquish the associated equipment capacity

♦ Apply technical knowledge and experience to the analysis of each plannable job

♦ Increase equipment availability

♦ Minimize downtime and interruptions to operations

Planning, coordination, and scheduling is a boon to Maintenance Supervisors when they:

♦ Define and measure workload, permitting advanced determination of staffing required in a given area or of a given skill

♦ Establish realistic priorities

♦ Identify the best methods and procedures

♦ Anticipate and preclude bottlenecks and interruptions

♦ Coordinate manpower, materials and equipment including:
  • Coordinated crafts
  • Parts and materials
  • Special tools and equipment
  • Shop and other support

- Off-site job preparation to minimize downtime
- Equipment access

♦ Preclude delays that would otherwise occur after work has begun due to waiting for information, materials, equipment, other skills, tools, etc.

♦ Establish expectations for the maintenance work force through provision of a detailed job schedule for the entire week, with individual time-lines as appropriate, thereby furnishing a control vehicle by which maintenance supervisors can monitor progress throughout the work day and work week

♦ Integrate preventive/predictive maintenance into the overall schedule

♦ Provide accurate promises that can be fulfilled

♦ Increase quality of output

♦ Control overtime

♦ Monitor job status

♦ Provides more time for direct job site leadership

Both Purchasing and Stores win when planning, coordination, and scheduling:

♦ Improve accountability for all parts and material

♦ Insure that parts are ordered with adequate lead time, reducing the number of emergency purchases and cost of express freight

♦ Optimize maintenance inventory

♦ Improve information available for equipment specification

EMPHASIZE THE MAINTENANCE DELIVERABLES

Maintenance is far more than a "necessary evil, " because it contributes in ways that are crucial to the success of the entity … as depicted in Figure 1.5.

Few drivers of the maintenance workload are discretionary. If maintenance is not funded and sized, employees are maimed or killed, our environment is destroyed, jobs are lost, management is fired, and the entity fails.

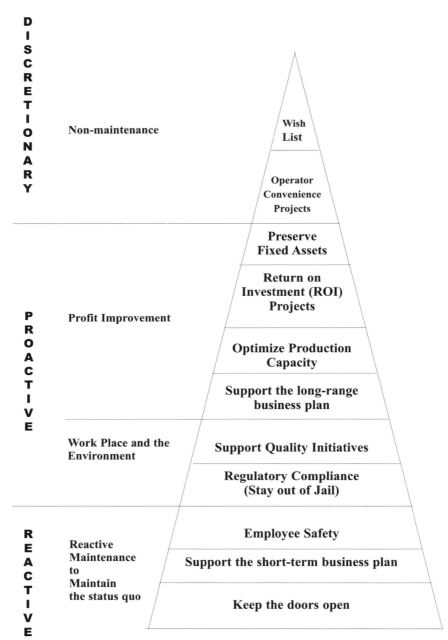

Figure 1.5 Build Up of The Maintenance Budget

Understanding the Nature of Maintenance Activities & Organizing Accordingly

Most maintenance departments are organized only for reaction to urgent demands, yet they wonder why they live in a reactive environment with a reactive culture. Organizationally, there must be recognition of and provision for the three broad types of work performed by the maintenance department: prompt emergency response, reliable routine service, and timely backlog relief. To be organized primarily for the first type (urgent response) is obviously detrimental to the latter two types.

To effectively support Operations with asset capacity and reliability, maintenance resources must be balanced with workload and distributed in a manner that assures effective performance of all three types of work. Deferral of critical maintenance leads only to further breakdowns and perpetuation of reactive maintenance; which leads ultimately only to failure of the business entity.

As managers, we must not allow urgency alone to consume all available resources. Resources must be preserved or provided for the important work that improves future reliability and thereby reduces future urgencies. While breakdowns and other suddenly developing conditions require prompt response, they are simply they urgent portion of the maintenance workload. Preoccupation with reactive response perpetuates status quo and causes the organizations future to look like its past failure to meet organizational objectives.

ORGANIZATION BY WORK TYPE

Effective control of the maintenance function depends upon clear accountability for each type of demand placed upon the organization. One form of maintenance organization is structured to facilitate control of each work type. While there are other structures that can be successfully deployed, this structure (Figure 2.1) best facilitates the understanding sought within this chapter.

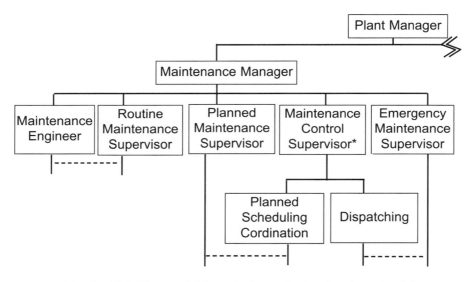

* Another title is Manager of Maintenance Support Services (see chapter 3 and 4)

Figure 2.1 Organization by Work Type

Such a structure is composed of three major operating groups covering the three principal types of demand. The concept is establishment of two minimally sized crews to meet routine and emergency demands with a third large group devoted to planned relief of maintenance backlog. Let us reach an understanding of the three groups and the work types they perform. Each group requires a different form of control.

❖ The Routine Group Provides Reliable Routine Service: This encompasses responsibility for performance of all management-approved routine tasks in accordance with detailed schedules and established quality levels. Most prevalent of this workload is the preventive/predictive maintenance process (PM/PdM). An optimized PM/PdM process, exclusive of major shutdown inspections, requires 15% to 20% of maintenance labor resources. Because the major inspections require a crew of several people to disassemble, reassemble, etc., they must be scheduled for performance by the large backlog relief group. Neither Emergency or Scheduled Backlog work should be allowed to interrupt the Routine Group; thereby protecting integrity of the PM/PdM schedule.

Routine work is specifically defined, of known and consistent content, performed in a defined manner, requires a predictable amount of time, and is performed according to a known schedule. It is pro-active and therefore should be planned, coordinated and scheduled. The planning should be thorough when the routine is established and further planning should not be required until the routine is re-engineered. Through the CMMIS, scheduling should be automatic when each routine falls due.

Routines include all PM/PdM and other inspections, as well as lubrication, calibration, tests, cleaning, adjustment, tightening, etc.

Routine crews carry tools and materials for small deficiencies commonly found. Time is provided for minor "corrective actions" of less than 20 minutes. Such actions are noted as part of PM/PdM feedback. Findings requiring more time trigger "Corrective Work Orders" which feed into the backlog and are planned and scheduled for performance by the backlog relief crew.

❖ The Emergency Group Provides Prompt Response to True Urgent Needs: It has responsibility for handling essentially all urgent demands; requesting assistance only when necessary. In other words, this group protects the other two groups from interruption. It cannot fulfill this objective 100% of the time, unless staffed for peak demand, which is not optimium. Therefore, they need to request help about 10% of the time. Because the Routine Crew should not be interrupted, assistance is provided by the Planned Backlog Relief Crew. In a pro-active maintenance environment, prompt response to urgent work requires approximately 10% of maintenance labor resources assuming the Emergency Group is staffed by multi-skilled personnel

Urgent work, by its nature offers little opportunity to plan or schedule except in a most rudimentary way. It should be radio-dispatched, mobilized, and well equipped.

❖ <u>The Planned Backlog Relief Group Provides Timely Relief of those Work Requests with Adequate Lead Time to be Planned:</u> Backlog consists of all plannable-work (non emergency or urgent) still open (fully or partially). As used in this book, it does not connote delinquency. In a pro-active environment, the bulk of maintenance workload should be plannable. Sixty-five to seventy-five percent (65% to 75%) of maintenance resources should work in this mode as part of the Planned Backlog Relief Group. Schedule Compliance for this group should exceed 90%. As stated previously, the Planned Group is called upon to support the Response Group whenever that group encounters peak demands (10% of the time by design). this still allows the Planned Group (constituting 75% of the maintenance workforce) to be scheduled and assigned to well prepared jobs and not be interrupted 90% of the time. This is far more pro-active than most maintenance departments anywhere.

Backlog work originates from PM/PdM inspections in the form of Corrective Work Orders, from projects (both capital and expense), and non-urgent requests from sources throughout the organization. Once backlog work orders have been properly prepare, they form the basis for coordinated Weekly Schedules (see Chapter 13).

The important portions of maintenance workload are reliable routine service and timely backlog relief. Proper attention here provides asset reliability, minimizes future emergencies, leads to a pro-active environment, and provides sustained viability of the organizational entity.

Sources of Planned Work	Percent of Total Planned Work
Results of PM/PdM inspections	30%
Scheduled component replacements	20%
Overhauls/rebuilds	15%
Internal Customer Input (Operators and Supervisors)	10%
Engineering project support	8%
Safety work	5%
Analysis of repair history	5%
Management directed work	4%
Service requests	2%
Accident damage	1%
Total Planned Work	100%

Figure 2.2 Sources of Planned Work

Regardless of organizational structure deployed, it must be equally responsive to the three work types. If not a large portion of the work force will be found in fixed or semi-fixed areas of assignment, in standby mode, waiting for the next emergency. Workforce utilization and asset reliability will be extremely poor. Planning and scheduling will be ineffective because few resources will be available to work on the jobs diligently prepared (planned, coordinated and scheduled). At best, Schedule Compliance will be only 50 to 60%.

Where Planning Fits Into Good Maintenance Practices

hen a maintenance planning and scheduling function is being established, the first question that usually arises is where and how it fits into the organization. The first answer is that it is structured within the maintenance organization, not outside it. Secondly, it should be organizationally independent of the specific maintenance supervisor(s) it is tasking, as well as supporting. This arrangement is based on the principle of checks and balances. If planning reports to first-line maintenance supervisors, there is tendency to establish lenient expectancies in terms of estimates and schedules. There is also a tendency to use the planner for daily expediting, clerical work and other reactive and inappropriate duties.

Planners therefore should be on the same organizational level as the supervisors they support on a week-to-week basis. Neither should be superior or subordinate to the other. Unless the maintenance department is large, both supervisor and planner should report to a common maintenance manager. If the department is large and has several support positions including planners, maintenance engineers, and administrators (clerks); a Manager of Maintenance Support Services position is recommended to lead all the support functions. In this scenario, the Manager of Maintenance Support Services (see appendix for a position description and Figure 2-1 for organizational posture) and the Maintenance Manager should report to a common Director.

The maintenance supervisor and maintenance planner form the most important partnership within the maintenance department. This relationship is sensitive, at least at the beginning. It is beholden upon planners to remember that they are staff in support of line supervision. Staff supports, counsels, recommends, and convinces line but does not issue orders to line. This rule must not be broken if the partnership is to prosper. Appendix F contains an article entitled "Recognizing the Pitfalls of Planning for Others." All planners should read and digest this article.

In terms of job grade, the planner position should be one grade higher than a junior first-line supervisor and one grade lower than a senior first-line supervisor. There is a need to attract experienced although junior, supervisors into planning and subsequently to attract experienced planners to senior supervisory positions as they pursue the career path to Manager of Maintenance Support Services or Maintenance Manager.

The Planner job is preferably salaried rather than hourly. Planning, coordination, and scheduling are managerial responsibilities. The first allegiance of planners must be to goals of the organization, not to the labor movement. Yet, the best maintenance planners come from the craft ranks. They have usually performed similar jobs themselves and can visualize the job being planned. Determining and estimating job steps and material requirements is thus easier.

Sooner or later, the question arises as to whether planners should be decentralized or centralized. Should their workstation be in the operating department(s) they are supporting or in a central maintenance location? The best answer to that question is that they need a desk in both places. Planning starts in the field at future job sites. It is best completed in a quiet location conducive to research and thought. Coordination and Scheduling are best done centrally where planner and customers can gather to resolve distribution of resources available next week in the interests of the overall enterprise.

Should Work Preparation be a Separate and Distinct Function?

The functions of planning, coordination and scheduling are essential to effective execution of maintenance work and thereby to achievement of maintenance excellence, asset reliability and world-class operations. As

Chapter 3

Peter Drucker once said "The productivity of work is not the responsibility of the worker but of the manager. A worker will not plan for his own efficiency." Craftsmen and supervisors are primarily concerned with getting jobs performed, not with preparation for their efficient execution. Dedicated, para-professional, managerial personnel best perform the three preparatory functions.

Unfortunately, under modern principles of human behavior, the common belief is that separation of preparation from execution is unnecessary. With arrival of lean, participative, organizational philosophies, staff support is being drastically reduced in favor of reliance upon "high involvement teams." These teams are expected to prepare their own work. Consistent with concepts of worker involvement and self-direction, job preparation is being forced to the lowest organizational level possible.

However, when mechanics are engaged in planning, they are not performing productive work. Furthermore, people preparing for their own work, one job at a time, are less effective than staff professionals specialized in the coordination of resources (labor, materials, parts, tools, equipment, technical documentation, support personnel, and transportation) required to perform a number of jobs in a given time period.

Furthermore, in today's high technology world, good maintenance mechanics are a precious resource. They are the primary resource aimed at preserving asset reliability and production capacity. In most unionized facilities, only mechanics are contractually authorized to use tools in performance of maintenance work. Yet, numerous studies (see Figure 1.2) show that maintenance mechanics spend, on average, only two or three hours per day applying the tools of their trades. It is imperative, therefore, that activities which consume mechanic time and are not related to tool usage be held to a minimum.

Participative worker involvement is a powerful vehicle for continuous improvement, but dedicated, structured, and focused preparatory support by well chosen and trained planners remains an essential requirement. Without such support, maintenance resources are applied ineffectively.

Unfortunately, many general managers are reluctant to acknowledge the requirement for dedicated planners. The options and related fallacies follow on the next page.

THE ASSIGNED CRAFTSMAN

If planning is left to technicians it is rarely performed well. It results in delay, wasted effort, and inefficiency. The worker engages in activities that reduce time available for direct work. Indirect activity and travel tend to be very high in proportion to direct work time.

Because of his position in the organizational hierarchy, the craftsman is not well-positioned for many of the liaisons associated with the planning and scheduling role.

THE RESPONSIBLE SUPERVISOR OR TEAM LEADER

Activities of first-line supervision should be focused on leadership, instruction, oversight, control and training in methods, workmanship, quality, feedback, and housekeeping. Given the demands of daily maintenance execution, supervisors are forced to concentrate on the immediacy of today's problems and have little time left to focus on effective preparation for future activities. If they are tasked to address both preparation and execution, planning for future jobs is almost always neglected due to the pressure of today's work.

Due to the demands of execution, supervisors typically plan just prior to the start of the job, thus leaving little time to consider methods and to identify and acquire required materials, tools, support crafts, etc. The result is improper crew size, missing items , ineffective methods, delays, unfinished jobs, and dissatisfied customers. Any time that supervision does spend on planning reduces the amount of time that would otherwise be devoted to field supervision and the many benefits thereof:

- Supervisory follow-through promotes better quality workmanship and improved productivity.
- Instruction and training develops apprentices and lower skilled personnel into "craftsmen" of the highest caliber—capable of maintaining reliability of today's high technology equipment.
- Good employee relations are fostered with reduction of grievances and minimal organizational damage through prompt handling of those grievances that do occur. The resultant high morale and team spirit contribute to increased output, efficiency and reliability.

Chapter 3

THE PROVEN ANSWER

Experience shows that the functions of preparation, supervision, and maintenance engineering are best separated. All three require different skills, and a combination of all these skills in one person is the exception rather than the rule. Therefore, planning should be a para-professional, staff function, separate from supervision and maintenance engineering.

Planning performed by a separate staff group has the added advantage of an overall functional perspective. Priorities, labor loading, and management reports are better coordinated through a structured function. A focused planner can plan several jobs more efficiently, than a craftsman planning part-time … one job at a time.

This separation of the planning function does not mean that the maintenance technician is not involved in the planning portion of preparation. Mechanics, supervisors and planners all contribute to the planning process. Craftsmen contribute in five important ways:

❖ First; some jobs either by design (minor) or default (urgency) are not covered by a planned job package. For these jobs, it is significantly preferable for the assigned mechanic to perform the work following planning forethought of his own rather than "charging" into it without benefit of any planning at all. Accordingly, it is beneficial to train mechanics in planning basics. With the basics ingrained, even emergency work orders gain some of the benefits of planning.

❖ Second; there are situations where the backlog waiting for planning gets out of control. This situation is inevitable unless the planning function is staffed for peaks, which is not desirable. Use of selected mechanics to "catch up" is recommended. Better to use a small portion of the mechanic staff as temporary planners than to lose control of the backlog and force the entire resource of mechanics to work without benefit of prior planning.

❖ Third; there are situations where the knowledge of a given mechanic simply exceeds that of the planners or supervisors. Such knowledge should be collected and committed to the electronic database for use by everyone. It is foolish not to draw upon the best knowledge available.

21

❖ Fourth; craftsmen can be utilized as planning assistants on a temporary, rotating basis. This approach is particularly appropriate when planner support spans are excessive due to budgetary limitations. A number of organizations are using this approach with outstanding results. Normally, the rotation period is three to six months. Four benefits are derived:

- Additional planning capacity increases planned job coverage and/or reduces the backlog waiting to be planned. This bottleneck is the most ludicrous of all backlog forms. Planners are infused into the organization to expedite the effective completion of requested work. Backlog should not be held pending planning. The longest period that jobs should sit in the planner queue is three days.

- Craftsmen gain perspective on the planning process and the benefits thereof. This insight improves their use of planned job routines and promotes improved feedback leading to improvement of the routines .

- Craftsmen return to their crew or team as disciples of the planning, coordination and scheduling process.

- Currency of the craftsmen's knowledge helps to keep the planner current relative to field conditions.

❖ A possible fifth involvement relates to participative team concepts whereby one member of the team is designated as the planner or coordinator for a period of time.

It should be remembered that craftsmen are responsible for the continuous and constructive feedback that is needed to optimize the effectiveness of maintenance planning.

CHANNELS OF COORDINATION AND COMMUNICATION

The Planning and Scheduling group is the hub of intra and inter functional/organizational coordination and communication. Planners are the principle point of contact and liaison between maintenance, operations and other supported departments. If they do their job well and are unbiased moderators between functions, supported departments feel that the planners are as much a part of their department as of the maintenance department. This relationship, called direct liaison is depicted in Figure 3.1.

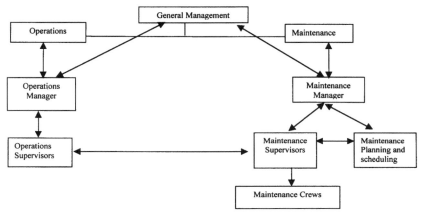

Figure 3.1 Direct liaison

There is two-way, closed loop communication within the maintenance department as well as between maintenance and internal customers. The primary linkage with customers is through planners, but this of course does not restrict direct communication between managers and supervisors and their operating counterparts.

The antithesis is presented in Figure 3.2, where planners are seen to support only the maintenance manager as a staff assistant. All other liai-

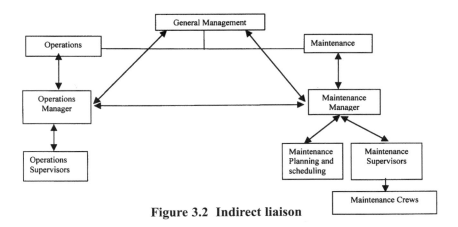

Figure 3.2 Indirect liaison

son is accomplished by line positions, as time allows, which, as we have already discussed, is not much. Your installation must not be allowed to go down this blind alley.

Working Liaisons

In large organization, in order to improve coordination between maintenance, operations and other internal customers, consideration should be given to the identification of primary points of contacts (Maintenance Liaisons). Within each major operational unit, such working liaisons are essential throughout the workweek, especially during the weekly coordination and scheduling process. Let's assume that there are six operating departments in a continuous operation requiring five shifts to cover 168 operating hours per week. Such a requirement implies 30 supervisors of operating crews placing demands upon maintenance. This is a lot of people for planners to coordinate with effectively. Therefore, it is recommended that large organizations establish a specific person in each operating department as a focal point for communication and liaison with maintenance. Such a person is frequently the process engineer assigned to the department, or perhaps an assistant superintendent.

Should Planning be Separate from Scheduling?

Planning and scheduling can be combined with one planner/scheduler supporting all planned work within an organizational unit (skill, area, or team) or they can be separated horizontally by specific activity (planning, material coordination, customer liaison, and scheduling). The selection is a local decision. Many organizations are too small to consider separation of the duties. They may have no more than a few planners and must then be structured as Planner/Schedulers (Figure 3.3). They might plan and schedule for one or more operating areas or for one or more maintenance skills. Responsibility by operating area is preferable, but planner knowledge limitations might dictate responsibility by maintenance skill. It is often difficult to identify a potential planner capable of planning electrical as well as mechanical work. This difficulty relates to the planning portion of the job, not to the scheduling portion.

24

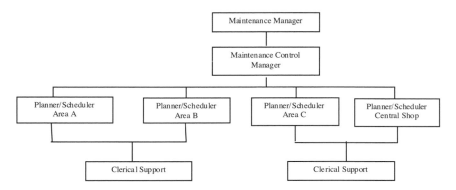

Figure 3.3 Horizontal Planner Organizatonal Structure

In large organizations with sufficient planner positions authorized, and incumbents with focused rather than general strengths, horizontal separation of planners, material coordinators, and schedulers should be considered (Figure 3-4).

If the functions are separated, planning is decentralized to where the work occurs, while scheduling is centralized with responsibility to distribute resources based upon location of the workload. Planners detail the work and identify material needs, and then hand off the job to Material Coordinators who source, procure, and expedite the needed materials. When all materials are on hand, jobs are passed to Schedulers who coordinate with operations and schedule maintenance resources to perform the job at a time when operations can best release the asset.

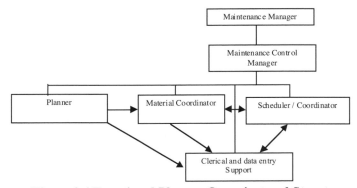

Figure 3.4 Functional Planner Organizatonal Structure

Selection of this structure is often based upon available skills. Planning requires more craft knowledge than scheduling. Separation of the two can conserve planning capability. Similarly, when the material management system is complex and system knowledge is limited, the material coordinator position should be considered.

Although more difficult in the latter arrangement, the Planner Scheduler control span should be held consistent to one planner/scheduler per twenty technicians (1:20). The split would be one material coordinator and one scheduling coordinator for every three planners (5/100 = 1/20).

A primary consideration in the decision (Fig. 3.3 or 3.4) is remoteness of the work centers at which maintenance takes place. If these centers are widespread, the latter structure is often chosen.

Another reason for separation is often capability of the available personnel. The best planners come off the shop floor, but these people often are not good time managers, so they may not make good schedulers. The best schedulers might come from Industrial Engineering or Production Scheduling, for instance. If some people have been inherited from the storeroom or purchasing, they may not be candidates for either planning or scheduling. However, they may be very strong in regard to sourcing, procuring, and expediting. They may be effective in the middle role (Materials Coordinator). Do not confuse the Materials Coordinator with the Coordinator/Scheduler.

In either scenario, planning activity needs to be conducted in the field where the work is to be performed, and scheduling activity is best conducted centrally so that resources can be allocated according to where the workload is heaviest from week to week.

CLARIFICATION OF ROLES

Duties that are not the responsibilities of planner/schedulers require as much clarification as those that are. The responsibilities of the three distinct functions that must be provided for in the management of maintenance are:

1. Supervision of work execution (supervisors or team leaders)
2. Engineering dedication to the elimination of repetitive failure (maintenance engineers).
3. Planned work preparation (planner/schedulers)

In **very** small departments the maintenance manager might perform all these functions, although probably not very well. As the department gets larger, each of these functions needs to be covered by an individual, and finally by several individuals.

The Maintenance Supervisor is responsible for the well being, training, and leadership of team members. The supervisor also has responsibility for control and follow up regarding:

- ♦ Proper, safe, and efficient execution of maintenance work
- ♦ Refinement and finalization of labor, parts, priorities, and methods, which are ultimately the supervisor's responsibility. Planner/ schedulers are staff support. The best of plans cannot anticipate all needs
- ♦ Job quality, duration, cost, and thoroughness of the completed job
- ♦ Time lost between jobs, at breaks and during shift changes
- ♦ Overruns and interruptions
- ♦ Having the next job always ready and assigned to a mechanic
- ♦ Tactical decisions necessary to remain as close to schedule as possible. Tactics often must be established after the job is started
- ♦ Assurance of reasonably accurate distribution of time and materials to specific jobs
- ♦ Communication with planner/schedulers, operations, and maintenance engineering as frequently as possible

In exercise of their responsibilities, Supervisors also must:

- ♦ Balance motivation and discipline by interfacing with each mechanic at least twice daily and visiting significant jobs two or three times daily
- ♦ Give adequate time and attention to formal and on-the-job training, never neglecting development of each team member. By this means, a Supervisor leads the training process; identifying, providing, or obtaining the skills training needed by each team member
- ♦ Act upon requests for support and provide prompt and fair handling of grievances. Effective listening reduces grievances.
- ♦ Control of tardiness, absenteeism, and vacations

The Maintenance Engineer is responsible for application of technical skills and ingenuity to the avoidance and correction of equipment problems causing excessive production downtime, quality variations, and maintenance work. This responsibility includes responsibility for:

- Maintainability of new installations
- Identification and correction of chronic, costly and dangerous equipment problems
- Technical advice to maintenance and proprietors (operations)
- Designing, monitoring, and continual refinement of an effective and economically justified preventive/predictive maintenance program, including:
  - Proper operation and care of equipment
  - Comprehensive lubrication program
  - Inspection, adjustments, parts, replacements, and overhauls, for selected equipment
  - Vibration and other predictive analyses
  - Protection from the environment
- Maintaining and analyzing equipment data and history records to predict maintenance needs

The Planner/Scheduler is responsible for logistical support to remove all avoidable barriers standing in the way of effective completion of maintenance work. The planner performs the bulk of the planning job before the maintenance work is started. This function encompasses:

- Customer liaison for all plannable maintenance work
- Job plans and estimates
- Assurance that all necessary logistics are identified and provided for
- Coordination of manpower, parts and materials, equipment, and access to the asset in preparation for effective work execution
- Cooperative and coordinated scheduling of jobs in the order of agreed-upon importance
- Full day of work, each day of the schedule week, for each man
- Arrangement for delivery of materials to the job sites
- Ensuring that even low-priority jobs are accomplished
- Continuous communication with all involved and impacted parties
- Maintains records, indexes, and charts
- Reports on performance versus goals

Control spans for the three functions relative to numbers of technicians, must comply with good maintenance practice.

Function	Control Span
Supervision	1:10
Planner/Scheduler	1:20
Maintenance Engineer	1:40

Figure 3.5   Control spans

These ratios assume there is adequate administrative/clerical, purchasing, and storeroom support. When the control spans are much wider than the above, none of the functions are performed well and maintenance effectiveness diminishes rapidly. In that event, the cultural transition from reactive to proactive maintenance with resultant asset reliability will probably not become reality.

FACTORS INFLUENCING PLANNER/SCHEDULER CONTROL SPAN

Control span ratios should be used only as guidelines. Planner staffing depends upon a number of factors and should be tailored to fit the specific local situation, considering:
- Number of craft personnel performing planned/scheduled work
- Current state of maintenance management organization
- Complexity of craft structure in use
- Level of planning and scheduling needed
- Method of estimating used
- Current state of planner support system (labor and material libraries, CMMIS, etc.)
- Complexity of operating organization supported by maintenance
- Level of liaison and coordination for which planners are responsible
- Structure of planning and scheduling organization
- Other maintenance staff support in place (maintenance engineers, maintenance clerks, material coordinators, training coordinators, PM coordinators, relief supervisors, maintenance control manager of maintenance support services)

- Degree of support provided by traditional Purchasing and Storeroom functions. If these functions are neglecting their traditional responsibilities, inappropriate duties fall to the planners including sourcing, procuring, expediting, receiving, stocking, picking, kitting, staging, securing, delivery
- Other peripheral responsibilities placed upon the planners

A Worksheet is appended (see page 191) to aid in determining the appropriate planner to craftsman ratio for a specific organization.

RELATIONSHIP WITH OTHER FUNCTIONS

To be effective, maintenance planners need:

- To be recognized as an important contributor to the maintenance mission
- Resources that are continuously balanced with the workload
- Clear definition of their relationships with maintenance superintendents, supervisors, mechanics and operations
- Work requests written by requestors with adequate identification, descriptive information, and sufficient lead-time to plan properly and schedule appropriate manpower
- Proper computer support to allow development of a comprehensive planning database
- Effective logistical support from purchasing and stores with timely, procurement, expediting, assembly, delivery, and availability of required materials and spare parts
- Effective storeroom support so that planners need only identify required withdrawals by preparing necessary stores requisitions, and need not do stock picking, job kitting, order staging, order security, or order delivery to the job site
- Effective purchasing support so that planners need only prepare purchase order requests, without need for the planner to source, purchase, prepare purchase orders, or track and expedite delivery
- Proper receiving support so that the planner is reliably alerted when purchased items are received
- Commitment from maintenance and operating management to hold structured weekly coordination and scheduling sessions to

establish priorities for daily, weekly, down day, and major outage work

- Adequate maintenance engineering support so that planners do not have to develop standard operating and safety procedures, and do not have to devote time or engineering attention to recurring maintenance problems
- Cooperation from maintenance supervisors, mechanics, and operating supervisors in the effective use and application of efforts put into meaningful planned job packages
- Feedback from mechanics and supervisors (maintenance and operations) regarding specific shortfalls in planned job packages so that improvement of future packages is facilitated
- Feedback from maintenance supervisors regarding compliance with and exceptions to the weekly schedule with noted "reason".
- Recognition that they are planners and not foremen
- A proper work station

As previously mentioned, when there are more than five positions in the control function (including planners, schedulers, material coordinators, clerks, dispatchers, and maintenance engineers), they should report to a position such as "Manager of Maintenance Support Services." Such a position brings to the group coordination, functional discipline, and integrity, as well as managerial acumen and clout. These benefits are not always gained by maintenance managers who are consumed by demands of supervision, budgetary control, and response to upper management. When responsibility for formal planning and scheduling are added, the burden often becomes too much for the maintenance manager to cover effectively.

The "Manager of Maintenance Support Services" position is discussed in further detail in Chapter 4 and a position description is provided among the appendices.

Managing the Planning and Scheduling Function

There are two issues to the management of planning. First is management of planning effectiveness–of planning as a function–and the second is management of the planners as people. Management of planning effectiveness involves measurements and benchmarking using prior periods and other similar information. The people management side of the equation will be described with approaches suitable for both large planning departments and small planning functions.

MANAGEMENT OF PLANNERS

Maintenance managers themselves, normally manage small planning groups of with one to three planners. In larger planning departments with several support positions (including planners, maintenance engineers, system administrators, clerks, etc.) a Manager of Maintenance Support Services is recommended. This person is responsible for supporting maintenance with a variety of services including planning, maintenance engineering, information suport (CMMIS), and occasionally the maintenance store room and procurement of maintenance needs.

Planning works best when the planners do not report to crew supervision. They should be on the same organizational level as the supervisors they support and not subordinate to them. If planners work directly for supervisors, the tendency is for them to be used as expeditors, clerks, or purchasing agents in support of daily activity rather than as planners for future (next week) jobs.

The Manager of Maintenance Support Services is the primary:

- Champion and Chief Advocate of the Maintenance Excellence process
- Marketer and Seller of Maintenance Control throughout the organization
- Manager of all maintenance support services (staff, not line)
- Chief planner/scheduler, chief of maintenance engineering, head librarian for the maintenance and planning technical library, and CMMS administrator
- Owner and Controller of the CMMS. As such he or she:
  - Controls system security
  - Steers and continuously seeks system improvements (coding, user friendliness, application, and reporting)
  - Trainer of all parties regarding their responsibilities to the system and their usage thereof
  - Enforcer of proper system input and usage by confrontation of the abusers
- Leader of Data Base, Job Plan, and Job Estimate refinement
- Quality Assurance and Control leader of CMMIS, Planning, Scheduling, and RCM (PM/PdM and Root Cause Analysis)
- Analyzer of system information and trends of improvement and maintenance metrics ... including balancing of resources with workload
- Developer and definer of the issues and recommendations stemming from the Maintenance Support Services Team
- Person who may be responsible for the maintenance store room and for purchasing
- Performer of special studies requested by management (or self generated)

MANAGING PLANNING

Managing the planning function usually boils down to regular assessment of continuous improvement. Is planning, coordination and scheduling paying off? Is product produced per maintenance dollar continually increasing? The answer had better be "Yes."

For details about measuring the effectiveness of maintenance and maintenance planning see Chapter 16 regarding Maintenance Metrics.

Backlog Management and Work Programs

C ontrol of the backlog is key to successful management of the maintenance function. Backlog is defined as the net workload, measured in labor hours, requested but not yet completed. If a job has been started, only the portion still to be completed remains in the backlog. Delinquent jobs are considered part of the backlog, but being in backlog does not automatically connote delinquency.

Maintenance work is continuously created over time as:

❏ Equipment is utilized
❏ Facilities are exposed to weather and intended usage (traffic)
❏ PM becomes due
❏ Processes are changed
❏ Management decides to make changes

Planning for Maintenance Excellence begins with "Macro-Planning", which is perpetual balancing of maintenance resources with maintenance workload. Backlog must be held within established, manageable, control limits by preserving a balance between resources and workload. If a process is not successful the organization cannot become and remain proactive.

Past misuse and deferral of essential maintenance work accelerates deterioration and thereby further increases the requirement for maintenance resources. When workload exceeds resources on an ongoing basis the maintenance organization falls behind the rate of deterioration and the amount of reactive maintenance grows. This situation is very common and very deleterious to morale, quality of products, safety, regulatory compliance, and quality of life for the maintenance worker. You can walk through a plant in this state and see the deferred jobs, broken and rusted items, and general disarray that result when deterioration gets the upper hand.

Once an organization fails to keep up with deterioration, it cannot maintain a proactive program if one is in place and certainly cannot climb out of an existing reactive state without an influx of adequate resources. Without such an influx it is impossible for the facility to be proactive because inspections both detract from the resources available for backlog relief and identify more corrective work to be added to the backlog. The resource-demand imbalance continually worsens.

As a practical matter, when an organization is in this state, any deficiencies found by proactive inspections are not likely to be corrected before failure. This condition renders the proactive crew impotent, and destroys the morale and effectiveness of the entire maintenance organization.

The current backlog borne by each crew or trade (measured in weeks) must be calculated and displayed on a monthly basis. The vehicle for the calculation is the "Maintenance Work Program" discussed later in this chapter. Smart management will use the work program to make decisions based on facts. Some of the decision areas include:

❏ Establishing required staffing by type of work and skill required.
❏ Maintaining required resources by adjusting overtime, contractor support or staffing.
❏ Creating a shutdown plan that provides adequate resources to complete the essential work within a condensed time frame.
❏ Evaluating capacity available to handle project work.
❏ Deciding if and when to utilize contract support to meet abnormally high demands.

Chapter 5

BACKLOG MANAGEMENT

Well-managed maintenance departments hold their backlog within established upper and lower control limits. Backlog ready to be scheduled is part of, but isolated within, the total backlog. Two to four weeks of "Ready" backlog and four to eight weeks of "Total" backlog are considered the norm.

Ready Backlog is just that, jobs that are ready to go. All tools, parts, materials, drawings, and authorizations are in hand. The job can go at any time. Total backlog includes ready backlog plus all the other open work orders for which something is missing. The job could be missing parts, authorization, budget, or some other element.

Too much backlog means excessive delay in response to customer needs and ultimately the dreaded **"D"** word – "Deferred Maintenance." Deterioration begins to gain the upper hand again and customer service suffers. People requesting lower priority jobs notice that their jobs never get scheduled. The resultant lack of customer satisfaction contributes to the low esteem in which the maintenance function is held in many organizations.

A well-designed and administered "Priority" coding structure with an aging feature can prevent indefinite delays to individual jobs in the backlog. Medium- and even low-priority jobs are still to be completed within a reasonable time frame.

When the 'Ready' backlog increases beyond 4 weeks or the 'Total' backlog exceeds 8 weeks the consequences can be severe. In a steel mill in Western Pennsylvania, the total backlog was 96 weeks! That meant that the workers would need almost 2 years to finish the backlog even if all production was halted. The consequence was a complete lack of meaningful control. In addition, there was low morale, a poor safety record, low profits, a nasty work environment and a near constant state of fatigue due to the constant pressure of being behind, and low self-esteem in the workforce.

Too little backlog indicates insufficient workload to keep the existing maintenance staff effectively deployed and utilized. Work tends to be stretched to fill out the day. A supervisor who has to scramble to find work will find his resources wasted.

Chapter 5

JOB STATUS

It is essential that planners know the current status of each job in the backlog. Such status is documented via "Status" code within the Computerized Maintenance Management System (CMMS). This code (Status) brings order out of chaos, and is the most important code in the entire system. An example of status code structure is presented in Table 5.1 below:

Code	Description	Form of backlog
P	Waiting to be planned	Total
E	With Engineering	Total
A	Awaiting approval	Total
PF	Pending Funding	Total
DF	Deferred	Total
PO	Waiting for PO to be issued	Total
M	Awaiting receipt of materials	Total
FP	Further planning required	Total
DP	Requires downtime – programmed shutdown	Total
PW	Requires downtime – weekend	Ready
DS	Requires downtime – asset not scheduled	Ready
DO	Requires downtime – window of opportunity	Ready
R	Ready to be scheduled	Ready
FI	Ready for fill-in assignment	Ready
S	On the current schedule	Not in backlog
CO	All Maintenance work completed	Not in backlog
CM	WO pending material closeout	Not in backlog
C	Closed to Equipment History	Not in backlog
CP	Print Revisions not yet received	Not in backlog

Figure 5.1   Work Order Status Codes

The Work Program's backlog is regularly compared to available labor resources and necessary adjustments (overtime, contracting, staffing) are recommended. Adjustments may require additions or deletions at one time or another.

The existence of a coherent backlog list will answer questions generated from all levels of management (and some workers). The CMMS must enable planners to access the backlog by:

- ❑ Planner responsible for the area, trade or job
- ❑ Job status (codes above)
- ❑ Crew/supervisor
- ❑ Asset, machine, or production line
- ❑ Originator
- ❑ Age of work order
- ❑ Due date

Each of the access paths to the backlog answers different questions from different questioners. For example, access of backlog by machine answers the question, "What other jobs should be performed while we have access to this unit?"

The Backlog must be complete, current, pure, and reliable. How would the integrity of your current backlog stand up to the following test (Figure 5.2)?

- ❑ Jobs that are completed, but nobody has bothered to close them out
- ❑ Duplicate jobs under different names
- ❑ Jobs over six months old?
- ❑ Jobs for which no one recognizes the originator or why the job was needed in the first place
- ❑ A poorly described job (no one can figure out what to do to what)
- ❑ Job status not filled in and nobody recalls what the status is. Were parts required? Were they ordered? Were they ever delivered, and if so, where are they now?
- ❑ Jobs that need to be done that are not listed in the backlog

---

Figure 5.2   Checklist for Backlog Integrity

The last item, "Jobs not listed in the backlog?" can be a special problem. Perhaps your customers know that you are already overloaded and ask why they should initiate any more requests? The reality is, if the work should be performed maintenance needs to know about it. It must be evaluated relative other demands with assignment of a priority ranking. An ignored maintenance problem might be catapulted to the forefront due to an accident, a production increase, an ergonomics push or a quality effort.

If the job is not added into the controlled backlog the organization does not officially know about it. The job simply sits in the background as part of the amorphous mass of undefined deferred maintenance swirling around. If a need exists, it should be identified and quantified as part of the backlog.

DEVELOPMENT OF WORK PROGRAMS

Work Programs are the vehicle whereby maintenance resources are perpetually balanced with maintenance workload. Without this balance, deferred maintenance increases progressively and the benefits of preventive/predictive maintenance (PPM) cannot be realized. If deferred maintenance is significant, equipment obviously is not in proper condition. PPM is not conceived to put equipment in proper condition but to keep it in that condition once it is achieved.

Work Programs also:

❑ Ensure that expectations for backlog relief are realistic
❑ Make allowance for all commitments to indirect activity
❑ Clarify the labor hours required for response to breakdowns and urgent jobs
❑ Clarify the labor hours required to meet PM/PdM requirements
❑ Define the labor hours of work to be loaded to each crew's weekly schedule
❑ Ensure that requested completion dates (real or implied by assigned priority) are met
❑ Ensure that even low priority jobs reach the schedule in a reasonable period of time
❑ Clarify the capacity to handle project work and thereby determine when to use contract support

Because maintenance is managed by controlling backlog within established limits, the current backlog upon maintenance crews (measured in weeks) must be calculated and analyzed. A Work Program should be developed for each maintenance team at least monthly. A weekly example follows on page 41.

WORK PROGRAM

Period Ending _____

CREW _____

AVAILABLE RESOURCES

Crew Size _____

Straight Time Man-Hour Available Per Week	800
Planned Overtime Per Week	96
Man-Hour Contracted or Borrowed Per Week	0
Total Man- Hours Available Per Week	**896**

LESS INDIRECT COMMITMENTS (Weekly Averages)

Lunch(if paid)	0
Vacation	120
Absence	24
Training	56
Meetings	40
Special Assignments	40
Average Man-Hours Loaned to other areas	40
Other Indirect	10
Total Indirect Man-Hours Projected Per Week	**330**
Total Man-Hours Per Week Available for Direct Work	**566**

Commitments Other Than Backlog Relief (Weekly Averages)

Emergency/Urgent (unschedulable)	100
Routine PPM	120
Other Fixed Routine Assignments	0
Sub Total	**220**
Net Resource Available for Backlog Relief	**346**

Backlog Data	Current	Backlog Weeks	Current	Target
Backlog Man-Hours in Ready Status	3200	Ready	9.2	2 to 4
Total Man-Hours of Backlog	4800	Total	13.9	4 to 6

Figure 5.3 Work Program

Work Programs consist of four sections:

- ❑ The top section quantifies Gross labor hours authorized each week including budgeted overtime and contract support.
- ❑ The second section quantifies resources committed to various indirect activities including vacation, absenteeism, training, meetings, etc. This sub total is subtracted from Gross resources to calculate the resources available during a typical week for direct work.
- ❑ The third section quantifies average weekly consumption of direct work resources for response to urgent conditions such as equipment failures causing production downtime. Included in this section are resources that must be committed to the timely performance of PPM routines.
- ❑ In the fourth section, the combination of the two consumptions (urgent work and PPM routines) are subtracted from the direct work total to quantify the weekly resources available for relief of the backlog.
- ❑ Within the fifth section, the resources available for backlog relief are divided into current levels of "Ready" and "Total" backlog to quantify backlog weeks compared to established control limits.

This data should then be plotted on a trend chart to trigger managerial action, as appropriate (see Figure 5.4). Resources should be flexed (adjusted) up or down as analysis indicates necessity.
There are three resources that can be flexed:

- ❑ Overtime is the easiest resource to adjust, assuming that it is not already excessive. An overtime range between 7% and 15% of paid straight time is recommended. Too little over time is not a sound maintenance strategy. Too much over time is not healthy, safe, or productive for either the company or the employees.
- ❑ Contract Support is the second most flexible strategy.
- ❑ Permanent Staff is the least flexible alternative, and should be chosen only when resources and workload are out of the specified range in terms of backlog weeks with no relief in sight, and when the condition appears to be semi-permanent.

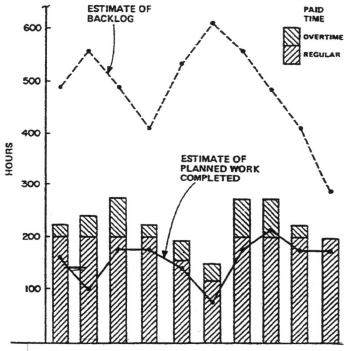

Figure 5.4 Backlog Weeks Trend Chart

Do not develop Work Programs only for the maintenance work force in total. This defines the average condition. Balance must be preserved at the crew level, not simply for the entire maintenance organization. The average condition for the entire organization might appear to be under control only because the backlogs of specific units, areas, skills or shifts are out of control (one or more to the high side, with others to the low side). The result is that the workload is not under control despite the overall adequacy of resources. Such an imbalance is common and the phenomenon suggests that although sufficient resources exist to get the job done, poor distribution of the available resources between the various organizational units (crews), where the workload exists, explains why backlog continues to grow further out of control.

Cross training and multiple skills greatly enhance managerial ability to distribute resources effectively.

Sizing the Maintenance Staff

Every year, at budget time, most organizations come under pressure from headquarters to reduce their maintenance budget. These requests prompt the question from all levels of management; how many maintenance positions are required to properly [1] maintain this facility? Although surveys offer some insight, the only meaningful response requires "definition and measurement of the inherent workload."

The staff required to preserve an asset is related to its size, replacement value, and usage. Without this investment the asset will **inevitably** deteriorate. Every shortfall below the required base staffing level that is not covered by increases in efficiency, effectiveness, or other forms of maintenance improvement will result in **deterioration**.

If reliance is placed only upon surveys or benchmarking to establish staffing levels, many unknowns remain. Some of the many factors that influence the required maintenance staffing are shown in Figure 6.1.

EXISTING STAFFING PROCESSES

Existing staffing processes for the maintenance function are generally inadequate and indefensible. Typically, maintenance does little to justify current staffing much less build a case for required staffing. At budget

---

(1) The word typically used is "adequately." Here, the authors have changed it, because ' adequately', has no place in World-Class Maintenance.

Figure 6.1   Contributors to maintenance cost and staffing

- ❑ Product and process
- ❑ Size of the facility, topography, and layout
- ❑ Access to the equipment for maintenance (how close together)
- ❑ Quality of design for the application
- ❑ Type of construction, materials and workmanship of buildings and equipment
- ❑ Age of buildings, equipment, and vehicles
- ❑ Whether the buildings are brand new or ancient or have historical significance
- ❑ Location of the unit, indoor or outdoor
- ❑ History of the site
- ❑ Amount and state of deferred maintenance today
- ❑ The expectation for maintenance?
- ❑ Small, medium, or large operations?
- ❑ The industry segment?
- ❑ The product mix?
- ❑ Single or multi-shift operation?
- ❑ Processes and equipment used
- ❑ Frequency of process improvements and changes
- ❑ Frequency of set up and product change-over
- ❑ Is purchasing based on life-cycle cost, or are purchases and contracts awarded to the low bidder?
- ❑ Policy and procedures for purchasing, engineering, and improvement projects.
- ❑ What percent up time, delivery, safety, and other measures of customer service are achieved
- ❑ How are the assets utilized
- ❑ Skill and dedication of the maintenance staff
- ❑ Knowledge, dedication, and expectations of the users (internal customers)
- ❑ Availability of spare parts and quality contractors in the plant locality
- ❑ Laws, building codes, and statutes that affect maintenance of the assets
- ❑ Size of the capital budget in relation to the overall asset value
- ❑ Amount of organizational change and turnover (maintenance, operations, and support)
- ❑ Competitiveness of the industry
- ❑ Hours of operation per week (1,2, or 3 shift, weekends)
- ❑ Production level, speed and automation of the operation
- ❑ ISO 900X
- ❑ Regulated industry (nuclear, pharmaceutical, etc.)

time the function is at the mercy of arbitrary reductions. But, without a basis on which to make a sound managerial decision, it is often difficult to identify which function is at fault (management, finance, or maintenance)?

Most organizations have a significant amount of trouble justifying additional resources; even when the facility has gained more equipment, more square footage and higher production levels than when current staffing was established. An effective work program system (described in Chapter 5) must be in place if the workload is to be defined. The work must be properly planned and measured to quantify labor hours requirements. Ultimately and ideally, this management information must be deployed to build a business-based argument for the required staffing.

Without records of backlog trends by craft there is no hard justification for additional resources. The maintenance leadership team is reduced to anecdotal evidence and indirect measures of increases in workload (rising customer complaints, increased downtime, growing use of contractors, etc.). Unfortunately such indirect measures at best are "lagging indicators;" they inform that problems exist well after much of the damage has been done. Sometimes a catastrophic failure is the main indicator that staffing is inadequate.

Like a business budget, a staffing budget can be built up from the demands your facility historically encounters. Figure 6.2 provides an illustration of how a business argument for the "required" maintenance staffing might be developed and presented to management.

Let's review the development from the bottom up, beginning with preventive/predictive inspections:

PREVENTIVE/PREDICTIVE MAINTENANCE INSPECTIONS

Presumably, the PPM inspections have been engineered with application of RCM concepts. The program has been presented and sold to management and the organization. In the example in Figure 6.2, the required staff resources were quantified at 15. Figure 6.2 demonstrates the build up.

Once management accepts a recommended PPM Program, these fifteen positions should be sacrosanct. The number should not be questioned without engineering reasons to modify the program. These resources will be applied to the detection of impending failures so that they can be corrected in a planned manner rather than under emergency conditions.

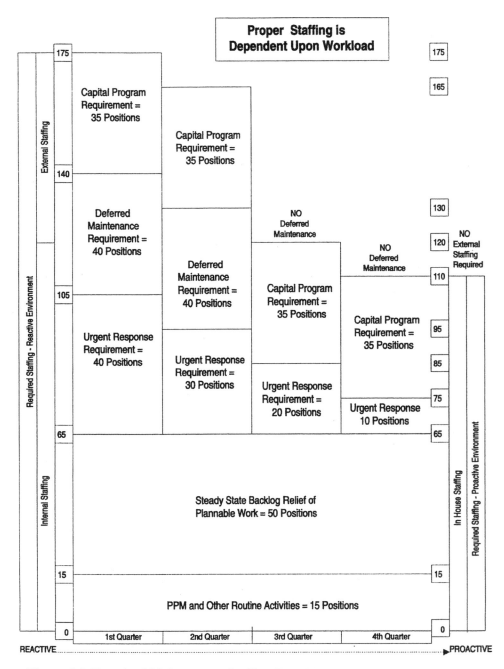

Figure 6.2 Required Maintenance Staffing Based Upon Workload – By Nature

Figure 6.3   Build Up of Required Staffing For PPM Inspections

Frequency of Routines	Labor Hours / Occurrence	Repetitions per Year	Labor Hours per Year
Daily	?	365	?
Weekly	?	52	?
Monthly	?	12	?
Quarterly	?	4	?
Semi-Annual	?	2	?
Annual	?	1	?
Less than annual	?	.5 to .1	?
		Total Per Year:	27,000 (assumed)
		Required Staff	15
		@ 1800 hours per year	

Do not develop this staffing number from history. History would provide the proper number only if on-time PPM Compliance was 100%. You need to know what resources are required, not what you might be able to invest ... given all the reactive demands upon your limited resources.

STEADY STATE BACKLOG RELIEF

The steady state backlog relief is the rate at which new requests for plannable work keep coming in. Please note that the 15 PM positions will be generating a good deal of this backlog through their inspection activity. Steady state does not include deferred maintenance, which we will come to later.

Go into the CMMS for the previous twelve months to establish a supportable number. Depending on your priority coding, you might be sorting on Priorities 3 and 4. If approved work and completed work differ, use approved or you may end up deferring approved work for lack of resources. Divide annual labor hours of new plannable workload by the average hours maintenance personnel work per year.

Let's assume the result of this analysis yields 90,000 hours, which justifies 50 positions. Refer again to figure 6.2.

Chapter 6

Urgent Response

Urgent response is the reactive workload that lack of reliability forces upon the maintenance function. Go back into the CMMS and sort by the emergency and urgent priority codes (perhaps Priorities 1 and 2). Divide the annual labor hours of the reactive workload by annual labor hours per position.

For our example, we have assumed the answer is 40 positions. However, last year PPM Schedule Compliance was poor. You didn't deploy anywhere near the required 27,000 man-hours developed above. But, in the future, you will have the required resources, so the need for reactive response should diminish significantly. This reduction is shown in the illustration as three decreases of ten positions each ... with an ultimate need of only 10 positions. That's the kind of return that can be expected (elimination of 30 reactive positions for an investment of 15 proactive positions).

Deferred Maintenance

The term-deferred maintenance encompasses all the requested and not disapproved work (it may not yet be approved or funded) that has not been completed in a reasonable time frame.

Go into the current backlog and summarize all jobs over three months old. Divide this total of hours required by the hours worked per person in a six-month period (six months being selected as a reasonable time frame to overcome the deferred state). The example assumes 40 positions required to achieve this objective. However, this provision is gone after mid-year. Once an organization achieves this worthy objective and experiences the benefits of proactive maintenance, it should never be allowed to slide back into the deferred state ... for that is false economy.

Capital Program Requirements

Translate the capital program for the coming year into terms of maintenance workers required. Our example illustration assumes a result of 35 positions.

Chapter 6

SUMMARY OF REQUIREMENTS

In summary the five analyses discussed quantify the total maintenance staff requirements for the coming year. The requirement is shown to be 180 at the beginning of the year with reduction to 110 by the end of the year.

The narrow second column from the left in Figure 6.2 suggests a proper internal/external scenario. Deferred maintenance and the capital program are not ongoing steady state situations. Organizations should be staffed for normal workloads, not for temporary peak-loads.

The example assumes current staffing of 125 positions with a natural attrition of 15 positions projected for the coming year. If half of the deferred maintenance is performed internally and normal attrition takes place, the 110 positions requirement at year-end can be achieved without forced layoffs.

Similar analysis and logic is the soundest methodology by which to properly size the maintenance work force.

OTHER CONSIDERATIONS FOR STAFFING

Each of the categories discussed above creates a demand for maintenance work and maintenance workers. There are four invisible demands that should be considered in the final decision. We call them invisible because they do not always show up on work orders.

Catastrophe demand: If your plant is located in a river valley you might have an infrequent flood, in the Snow Belt the occasional blizzard, and the (hopefully) rare fire for everyone. The point is that small catastrophes actually happen regularly. The question is: should provision for a blizzard or small fire be factored into maintenance staffing?

Construction related demand: Major construction on site will cause disruption to the maintenance schedule. Tradespeople are pulled off jobs to escort contractors around the facility, adjacent areas require extra cleaning, utilities are disrupted and other problems occur. When planning a large construction project, degradation in maintenance efficiency is to be expected and should be reflected in maintenance staffing.

Social demand: The maintenance demand created by visitors such as regular tours and irregular visits from company brass and outside VIPs may be termed social demand. Imagine what will happen to the maintenance schedule a week before a US Presidential visit. In some facilities, such an event is not a problem and in others significant time is spent on such activities.

Personal service demand: Maintenance people are sometimes used off schedule for personal service such as package delivery, airport pick-ups, home remodeling, or working on company sponsored community projects (such as building a playground). Much of this kind of work is off the work order system.

For all these demands, evaluate the impact on the ongoing operation. The best practice is to always use the work order system so that such additional items are accounted for and their needs can be projected.

ANOTHER APPROACH TO THE STAFFING QUESTION

An academic formula does exist for the determination of maintenance staffing. This formula provides a quick and approximate method of calculating staffing for a new or existing operation:

$$MBS = \frac{[(RCB \times BMR) + (RCE \times EMR)] \times NMF \times LP}{AASTLR}$$

Where:

- ❏ **MBS** = **Maintenance Budgeted Staffing**
- ❏ **RCB** = **Replacement Cost of Buildings** – Stationary Equipment such as building utilities are included. R.S. Means Costing Information may be helpful if this value is not readily available.
- ❏ **BMR** = **Building Maintenance Ratio** - The percentage of asset value which must be reinvested to ward off deterioration. For standard building types the common range is 0.5% to 2% per year. A 2% BMR infers that the entire building would be replaced on a 50 year cycle. The exact amount depends on the issues brought up at the beginning of the chapter (including building use, climate, and type of construction.)
- ❏ **RCE** = **Replacement Cost of Equipment** - This factor includes process, production, packaging, warehouse, and mobile equipment. In capital-intensive industries the RCE is of significantly greater magnitude than the RCB.

❑ **EMR** = **Equipment Maintenance Ratio** – For much equipment this ratio is 3% to 15% per year of RCE. One substantiation for these numbers is that service contracts for electronic equipment, including full labor and parts, are often priced at 12% of replacement cost (which includes both profit and travel).

❑ **NMF** = **Non Maintenance Factor** – The relationship between total time and material spent on all activity performed by the maintenance function and the time and materials spent on true maintenance, as opposed to non-maintenance work which the maintenance department performs (renovations, new construction, operating support such as set ups, change-over, and replacement of production expendables). This factor must be included in the equation. Although they are commonly among the responsibilities of the maintenance department, these efforts are not required to preserve the assets and therefore are not, by definition, true "maintenance."

❑ **LP** = **Labor Percentage** - This value is the percentage of the total maintenance budget that is "Direct Labor" without fringes or over-time premiums. If these two latter items are included in LP, they must also be reflected in the denominator of the equation (AASTLR).

❑ **ASSTLR** = **Average Annual Straight Time Labor Rate** – This face tor is the average hourly labor rate of all maintenance employees annualized.

If there are several large categories of assets, consider applying sep-arate factors for the different categories (i.e., production equipment versus mobile fleet equipment verses buildings).

Example of the Calculation

Using the same simulation as conveyed in Exhibit 6.2, the following information might be applied:

❑ RCB   = $ 50,000,000
❑ RCE   = $160,000,000
❑ BMR   = 2%
❑ EMR   = 7%
❑ NMF   = 1.25
❑ LP   = 50%
❑ ASSTLR = $ 41,600

In applying the formula we find:

MBS $\quad = \dfrac{\text{(RCB x BMR) + (RCSE x EMR) x NMF x LP}}{\text{AASTLR}}$

MBS $\quad = \underline{\text{(\$50MM x .02) + (\$160MM x .07) x 1.25 x .50}}$

$\qquad = \dfrac{\text{(\$1MM + \$11.2MM) x 0.625}}{\text{(2080 Hours/Year x \$20/Hour)}} = \dfrac{\text{\$7.63MM}}{\text{\$41,600}}$

MBS = 183 Positions

    This formula makes no provision for custodial services, and is intended only as an estimate, as it may be off by 30%, plus or minus. However, it would be helpful in answering a general question such as "How many maintenance positions are required for the new process?"

7

The Planning Process

(Micro-Planning)

The process of job planning encompasses verification of all aspects of the job to be done as well as identification of the various input resources (material, manpower, and equipment) required to complete each job in an orderly manner and at optimal overall cost. Despite being the key to maintenance effectiveness, "planning" has different meanings for different people, depending on background, experience, and application.

The best understanding can be achieved by establishing the criteria of a planned job, as illustrated in Figure 7.1.

STEPS OF THE PLANNING PROCESS

To develop a beneficial job plan requires that a logical, step-by-step process be followed. These steps correspond to the criteria of a planned job referenced above.

After defining and describing the job objectives and scope, list the steps to be performed; thus defining "what" is to be done. Determine if the job or a substantially similar job has been done before. You may have a job plan already on the shelf. If the job is not familiar to the planner, a visit to the job site and a discussion with the requestor is the next step. Pinning down exactly what is to be done is sometimes half the battle.

Chapter 7

Figure 7.1 Criteria of a planned job

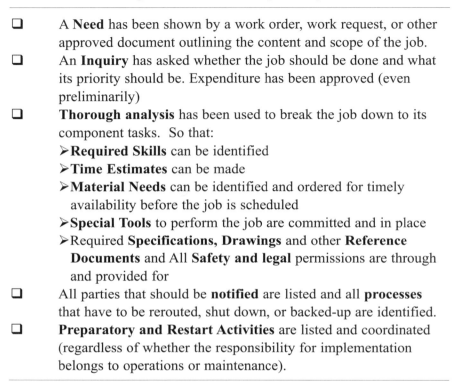

□   A **Need** has been shown by a work order, work request, or other approved document outlining the content and scope of the job.

□   An **Inquiry** has asked whether the job should be done and what its priority should be. Expenditure has been approved (even preliminarily)

□   **Thorough analysis** has been used to break the job down to its component tasks. So that:
  ➢ **Required Skills** can be identified
  ➢ **Time Estimates** can be made
  ➢ **Material Needs** can be identified and ordered for timely availability before the job is scheduled
  ➢ **Special Tools** to perform the job are committed and in place
  ➢ Required **Specifications, Drawings** and other **Reference Documents** and All **Safety and legal** permissions are through and provided for

□   All parties that should be **notified** are listed and all **processes** that have to be rerouted, shut down, or backed-up are identified.

□   **Preparatory and Restart Activities** are listed and coordinated (regardless of whether the responsibility for implementation belongs to operations or maintenance).

A preliminary go-ahead is usually given at some point early in the process. If the job is a routine repair (even a large one) the level of approval might be casual. If the job is new, special, or outside the regular business, approval might be very formal, with top management signatures required. Summarize the required budget for the job, thus defining *"how much"* it will cost and obtain whatever authorizations are required.

Taking the physical location and spaces around the equipment into consideration, plan the manner in which the work is to be accomplished, thus defining *"how"* the job will be performed. Include ideas about how many people can work efficiently and consider the movement of materials into and out of the area.

Establish duration and manpower needs required to *"perform"* the work. Determine the skill sets needed, and any special licensing requirements (i.e., certified welder). Clarify the sequence of skills required throughout job performance.

Will the work be done in-house, will a contractor be called in, or is some combination of the two required to do the job? These steps define *"who"* is to do the *"what."*

Next, identify and arrange for delivery of all spare parts, materials, consumables, special tools, PPE (personal protective equipment) and equipment necessary to do the job. Determine if the items are in stock. If so, reserve them. If not, determine where and when they can be obtained. The latter step should be accomplished with the support of Purchasing.

Then, determine essential reference materials and include them in the Planned Job Package.

Finally, planning is not completed until everyone knows what is going to happen. *Communicate* to all parties involved as well as to management, the plan of WHO is to do WHAT, HOW, by WHEN and for HOW MUCH.

The when step moves you out of planning and into coordination and scheduling. Coordinate with the asset custodian to select a mutual time when maintenance can have the resources available and operations can release the asset. Schedule the work; thus defining *"when"* the job will be performed.

When all these steps have been accomplished, a job plan has been prepared with a schedule for execution according to priorities and equipment availability established with Operations. Supervisors are relieved of preparatory activity, so that they can devote effort to leading job execution. Craftsmen will be far more productive because delays and conflicts are resolved on paper rather than on the shop floor.

WHAT WORK ORDERS SHOULD BE PLANNED AND HOW MUCH PLANNING IS ENOUGH?

The appropriate degree or detail of planning is often questioned. The question "which jobs should be planned?" applies primarily in early installation phases of planning and scheduling when there is insufficient capacity to effectively plan all jobs. The level of detail is less during early phases of implementation than in latter phases.

➢ Generally, larger jobs are planned first on the theory that larger jobs are usually accompanied by delays and conflicts and therefore there is greater opportunity for benefit from planning.

Surprisingly, much benefit can come from planning shorter jobs too.

➤ Cut-offs of four or eight hours are often established for the magnitude of jobs to be planned. While these limits may be acceptable in early phases, the selected cut-off point should be progressively reduced as the planning process matures.

Conversely, delays encountered in smaller jobs have a more dramatic percentage impact, and planning coverage should therefore ultimately include all jobs that can benefit. Although detailed planning can be more effort than justified on simple jobs, the usual tendency is to under-plan large jobs rather than to over-plan small jobs.

➤ Common thinking is that smaller jobs require less planning. However, even a one-hour job with essential materials missing can cause considerable loss of time due to unneces sary travel. A shorter job will have greater lost and unproductive time as a percentage of total time than a longer job.

➤ The preferred approach is to focus early efforts on the more **repetitive jobs**. By this means, Planner workload is reduced as the library of planned job packages grows. This benefit is a prime reason to provide ample planning capacity during early phases of program installation.

➤ Consider application of the Pareto Principal during early phases of program installation, 80 % of the benefit is commonly derived from 20 % of the effort.

When sufficient planning capacity exists, "all jobs that can benefit" should be planned. To achieve this end, planners must avoid getting involved in emergency work. Planning thoroughness increases as installations mature and varies with job repetitiveness. If the job is repetitive in nature, the Planner can afford to invest more time because the efforts will yield residual returns in months and years to come. When jobs are planned in "building blocks", previous planning effort is often applicable as portions of new job packages.

Eventually, planning should cover 80% or more of maintenance man-hours. The limited planning associated with the other 20% (emergency and unplanned for one reason or another) falls upon the supervisor and his team to execute effectively with as much forethought as possible. Supervisors and mechanics can also be used to increase planning coverage as necessary.

❑ An established Maintenance Plan for each asset … developed through the RCM process
❑ Timely reporting of potential problems by operations (to provide lead time for planning)
❑ An up to date Maintenance Technical Library (see below)
❑ An extensive library of Planned Job Packages
❑ Good use of a CMMIS
❑ Meaningful feedback on completed jobs by supervisors and technicians
❑ Complete equipment repair history
❑ Thorough PM/PdM inspections
❑ Thorough failure analysis
❑ Good relationships with vendors
❑ Open dialogue with operators and engineering on troublesome equipment
❑ Existence of overhaul and rebuild capabilities
❑ Good workmanship by craft personnel
❑ Good use of repair technology

Figure 7.2 Conditions that support planning effort, how many can you check off?

THE MAINTENANCE TECHNICAL LIBRARY PROVIDES INFORMATIONAL SUPPORT FOR THE PLANNING EFFORT

The maintenance technical library (MTL) is a place where maintenance planners (among others) have access to a wide variety of maintenance information including equipment history, equipment manuals, parts lists, and assembly drawings. Ideally the library should have Internet access, large reference tables, shelving for books and catalogs, and legal size file cabinets.

In addition the MTL should be the location of copies of plant drawings, site drawings, vendor catalogs, handbooks, engineering textbooks, and so on. . Access to software systems (CMMS, CAD/CAM, etc.) should also be available in the library area.

Considerations:

1. Protect the paper and computer records from disaster such as fire, flood, and theft. Consider fireproof file cabinets and off-site storage of copies.
2. Use some kind of sign-out system if material must be removed from the area.
3. Make it someone's responsibility to keep the records up to date (possible job for clerk or data entry person) and coordinate this effort with ISO 900X certification and regulatory requirements.
4. Manage the revisions so that all copies are updated (again coordinate with ISO 900X).
5. There are computer programs (document management systems) that allow the storage and retrieval of all these documents from any computer on the network (eliminating the need for a physical place for the library). Modern CMMS have document management capabilities.

Once the documents are cataloged and access to the computer network and the Internet is arranged, and fully operational, the benefits for the planner of a well-organized library are:

❑ The planner's job is simplified and accelerated
❑ Job plans are of consistent quality
❑ The library becomes a universal resource for maintenance engineering, specification, and problem solving
❑ There is a good foundation for further computer assistance

Good records in these areas are important and should include:

Equipment Records containing all pertinent data for equipment such as installation data, make and model, serial number, vendor capacity, and so on. An important item is original capacity or specification (10 tons per hour, 4M feet/min, etc) of the asset. Equipment records should not be confused with Equipment History of repairs made to the equipment.

Equipment Histories capturing all work order history performed against each equipment number.

Prints, Drawings and Sketches as installed.

Libraries (Planning Aids): Some planning functions vary each time they are carried out. Others follow the same pattern for each group of identical machines. It is therefore possible to simplify some of the planning processes for machine repair and overhaul by *classifying* identical groups of machinery and then *building* libraries of preplanned work element sequences and bills of materials for each class. (See chapter 11-Analytical Estimating).

The basic concept of these libraries is to establish and document the work sequences needed for each type of equipment, class by class. The documentation should record the procedure needed to take the equipment completely apart and then put it back together with replacement parts as needed.

Labor Libraries These libraries should be filed in some manner designed for convenient retrieval, normally by unit of equipment, specific type of skill required, or by job code. The labor library supports development of job step sequence and labor resource requirements, listing:

> ➤ Crew size by skill
> ➤ Job Duration
> ➤ Man Hours by skill

Labor Estimating System This system should provide basic data for building job estimates. Even where a labor library is in place, some form of estimating system is required to extend it as necessary.

Material Libraries This library should contain parts list, and bills of material filed by each unit of equipment on which maintenance work is performed. The material library supports identification of material/parts requirements for jobs to be planned, including:

> ➢ The parts involved
> ➢ The stockroom number of each part
> ➢ The manufacturer's I.D. for each part
> ➢ Storeroom location
> ➢ Unit of issue

Purchasing/Stores Catalogs These catalogs (electronic and hard copy) are essential for all parties. They contain much the same information as the material libraries used by planners. Indeed, some form of stores catalog or vendor catalog were used to develop material libraries. These catalogs should be sorted by:

> ➢ Component description
> ➢ Asset(s) where used
> ➢ Stock Number
> ➢ Vendor serial number

Files of Planned Job Packages containing previously developed packages for repetitive jobs to enable repeat usage with minimal duplication of planner effort.

Standard Operating Procedures are to be included in planning packages without repetitive documentation effort. Such procedures include safety, lockout, troubleshooting sequences, etc.

Other reference sources Reference sources should be utilized as the situation dictates:

- Service Manuals
- Planner Experience
- Supervisory Experience
- Mechanics
- Maintenance Engineers
- Operators

Chapter 7

THE PLANNED JOB PACKAGE

The planned job package for any given job contains documentation of all planning efforts performed for that job. Any factors that may delay or hinder effective job completion should be anticipated and provided for in the planned job package.

As appropriate, the assembled package is reviewed with the Maintenance Supervisor and the Requestor. Given the data contained within the package, coupled with a thorough verbal exchange between the planner and supervisor and a similar exchange between supervisor and mechanic, nothing should be lost between strategic planning and tactical execution.

The Planner then holds the Planned Job Package for necessary procurements.

Figure 7.3   A complete planning package may include:

Work order
Work planning sheet
Job plan with details by task with step-by-step procedures
Labor deployment plan by craft and skill including labor-hour estimates. Consider contract as well as in-house resources. If the job is complex, consider the use of the GANTT bar chart or PERT network chart to help plan task sequencing to assigned crews. Maximize pre-shutdown fabrication and other preparation to minimize time the asset is out of service.
Bill of Material. List all materials needed for the job, including an acquisition plan for major items. Determine if the material is authorized inventory or a direct purchase item. The planning package should include spares reservation and staged location.
Requisition and purchase order reference list
Time for each step (task), summarized by resource group and for the total job
Sight set down plan (where to put everything used for major tear downs)
A copy of all required permits, clearances and tag outs
Prints, sketches, Polaroid/digital pictures, special procedures, specifications, sizes, tolerances and other references that the assigned crew is likely to need

Chapter 7

This overall micro-planning process is discussed in detail within the next five chapters:

Chapter 8 – Screening, Scoping, Research, Detailing, Permitting and Assembly of the Job Plan Package

Chapter 9 – Determining and Procuring Spares, Materials, Tools and Equipment

Chapter 10 & 11 – Work Measurement

Chapter 12 – Coordination with Operations

The Planning Process Screening, Scoping, Research, and Detailed Planning

Job Planning is a six-step process (plus a seventh step of feedback) that includes:

❑ Screening of Work Requests

❑ Assessing and Scoping the job to be performed

❑ Job Research to avoid redundant planner effort

❑ Job Breakdown with Detailing and Sequencing of Job Steps

❑ Material Take-offs and Procurements (Chapter 9)

❑ Assembling the Planned Job Package

❑ Receipt of feedback and reflecting it in updated job plans (Chapter 15)

SCREENING WORK REQUESTS

Work streams into the planner's in-box from many sources including PM inspector reports and customer requests. With support of Maintenance Administer/Clerks, planners review all work requests except those that must be performed on the same day as requested. Such requests for immediate action are handled directly by the Maintenance Supervisor without benefit of Planner support.

On Work Order Requests (WOR) that are to be planned, the Planner reviews and screens each for redundancy, necessity, completeness, and accuracy. Within this review, the Planner confirms that:

❑ The request is not a duplicate. Sometimes this is a particularly difficult task when the backlog is large. A major reason for this difficulty is that the same job may be requested using very different descriptions. For example, 'Remove and Replace a particular pipe' might be the same job as 'Correct seepage into sump.' It takes a sharp eye and good knowledge of the plant to spot likely duplicates. Effective CMMS's contain helpful features in this regard.

❑ The description is clear and complete with the appropriate Equipment Code. When work requests are reviewed it is often found that the description is inadequate and/or equipment code is wrong or not even entered. Explanations go something like: "everyone knows what we mean by the hissing spider"- (the shop floor name for a case sealer that has a bunch of hoses and hisses when in operation). Accurate descriptions and specific identification of the involved asset make everyone's job easier. Of course, this is why a well-designed equipment numbering system is vital. The numbering scheme must be readily available to all users. If they are initiating a request while on the operating floor, the numbers should be stenciled on the equipment it self. If initiating while in their office, they should have a pocket catalog of their equipment numbers for easy reference.

❑ All requestor required fields are completed with valid codes. Effective analysis of Equipment History is dependant upon good descriptions and proper coding.

❑ Priority and requested completion dates are realistic and provide practical lead-time, enabling the job to be prepared for effective execution.

❑ Authorization has been given. Different types and sizes of jobs require different types and levels of justification to obtain authorization. The planner may have to develop a preliminary estimate on which to obtain approval. Engineering approval should be a requirement for all alteration and modification requests (before maintenance processes the request any further). Some of the worst industrial accidents in the world have resulted from field modifications that were undertaken without the benefit of formal engineering review.

❑ The requested work is needed. If so, does it need to be accomplished at this time? In this sense the planner serves as a gate-keeper of maintenance resources. If questioned about the need or priority, the issue is resolved with the requesting department or referred to the Maintenance Superintendent for resolution.

ASSESSING AND SCOPING THE JOB

Before progressing further, the required level of planning must be determined. Does the job warrant detailed planning or only cursory planning? Is the effort and cost worth the value to be gained? This issue relates essentially to the early phases of a Planning installation, as there is seldom-sufficient capacity to plan all jobs at the outset. However, as the library of previously planned job packages expands, and the planner workload becomes manageable, all jobs that can benefit should ultimately be planned.

One-third of the Planner's day should be spent visiting job sites to analyze jobs to be planned. The experienced planner is always striving to minimize the number of unknowns, which dramatically increase time required to complete jobs. Planners always guard against "assumptions," that often prove to be wrong, with serious adverse impact on Schedule Compliance. The best way to catch such wrongful assumptions is to get out of the office and visit future job sites before trying to plan the jobs, and to visit active job sites to learn how job packages might be improved.

The process of job assessment and scoping should roughly follow the flow outlined on the next page.

Figure 8.1   Job Assessment and Scoping Checklist

❑ Confer with the requestor to clarify the desired result. By clarifying the end objective, the means to get there can often be simplified or the job can be expanded to solve several problems at once.

❑ Refine the description accordingly

❑ Clarify the specific job location (building, floor, bay, etc.)

❑ List what needs to be done (job content)

❑ Define start and finish points (job scope)

❑ Finalize priority

❑ Visualize job execution and outline the requirements

❑ Record the steps necessary to execute the job

❑ Prepare sketches or take photos to clarify the intent of the work order for assigned mechanics or simply as a reference during detailed planning

❑ Take minor measurements (exactly). Complex measurements (i.e., for fabrication) should be left to the assigned technicians.

❑ Determine required conditions. Must this job be coordinated with Operations?

❑ Is it necessary for equipment to be down (major or minor?)?

❑ Define involved control loops

❑ Determine if other equipment or adjacent areas will be impacted by performance of this job?

❑ Check for safety hazards

DEALING WITH SCOPE CREEP

Properly defined, scope means "the range of one's perceptions, thoughts, or actions," and its use is crucial to delivery performance and schedule compliance. Scope creep needs to be built into the job plan because some jobs are likely to creep. Let's consider a typical do-it-yourself scenario:

A young couple decided to replace a sink base cabinet in a mother-in-law suite soon to be occupied by a live-in babysitter. They assumed the job could be done over a weekend for about $350 worth of materials.

They pulled out the cabinet and noticed the wall behind had never been installed correctly, so they ripped that out. With the wall removed they had the opportunity to pull wires for new electrical circuits and to add a GFI. At this point they noticed that a small change would make the kitchen more efficient. So the gas pipe to the range was moved to accommodate the new set up.

The wall was reconstructed and received a nice paint job. After the cabinet, range, and refrigerator were reinstalled the rest of the room looked shabby. So the remaining walls needed a fresh paint job. Now the range looked old and needed to be replaced. Presumably, the job was finished, except that the floor covering looked out-of-date and worn.

Eight weekends and $1,750 later, the new babysitter was able to move in. This story illustrates how job scopes "creep" without proper planning and definition, pricing, authorization, and budgetary control.

The remainder of the planning cycle is completed at the planner workstation (taking two-thirds of the planner's day).

JOB RESEARCH

The third step is to search through labor libraries, reference files and the Maintenance Technical Library to determine if the job or portions thereof have been previously planned and to fill in information, knowledge and reference gaps. Always use what you can to avoid redundant effort.

Within the function of job research the Planner looks at several things. These tasks save the company money, save the planner time, shorten out of service time, and may even save a workers life. During Research Planners should:

❑ Use Labor Libraries and Equipment History to determine if the job has been previously performed. When was it previously done? Is this frequency excessive? If so, consider if anything can be done to avoid recurrence. Always ask, "Is this the best solution to the problem?" If the planner doesn't ask these questions, who will?

❑ Consider alternative approaches. Should additional work be performed this time to ensure a more permanent solution? Should the item be replaced rather than repaired (the Repair/Replace decision)? Should the item be purchased rather than made (the Make/Buy decision)?

❑ Remain conscious of alternate plans for the involved equipment. If there are plans in next year's capital budget to replace this equipment, why perform a major overhaul at this time?

❑ Consult the procedures file to identify necessary tag outs, safety inspections, fire watches, and standby positions associated with open flames, ladder and scaffold use, vessel entry, etc. Safety must always be a primary concern within the planning and execution processes. Planning is critical to safety and loss prevention, because the planning stage is the logical time to think about and incorporate safety matters into the Planned Job Packages, before the pressure to finish jobs comes into play.

❑ Contact other functions with involvement or potential input in the job. Planners are not necessarily engineers, supervisors or craftspersons, but they must use judgment and know when to call in the specialists from engineering, operating, safety, process control environmental and quality, functions, as well as appropriate contractors.

DETAILED JOB PLANNING
(DETERMINING THE BREAKDOWN OF JOB STEPS)

When research is complete the Planner prepares details and phases of the job requirements. These efforts are core to the planning process; describing a job in terms of how it is to be performed and what resources will be required. The Planner examines the job to be performed and determines the best way to accomplish the work. He must know the job well enough to describe what is to be accomplished and to estimate how many man-hours will be required. If the Planner doesn't know and record the requirements, the assigned crew will not know the expectation.

Within the process, the Planner must:

❑ Select and describe the best method for job performance

❑ Determine and sequence the job by specific and logical tasks or steps

❑ Identify task dependencies and consider application of PERT or CPM network analysis to facilitate the planning of complex jobs (see Chapter 18).

❑ Determine labor resource requirements including required skill sets for each task (craft and skill level). This step may go beyond mere craft identification to detail of skill sets). The planner determines the required crew size and labor-hours for each task of the job sequence, by applying available benchmarks (whatever is available at that time in that facility). After the work is accounted for, the planner applies normal job preparation, travel, and PF&D allowances (Personal, Fatigue and Unavoidable Delay) and determines whether special or extra allowances are required for the specific job being planned.

❑ List determinable materials, parts, and special tools required and prepare the Bill of Materials for the job. In consultation with Purchasing, establish the acquisition plan. From the CMMS inventory module or inquiry to the Storeroom, determine what items are in stock, reserve them, and prepare associated Stock Requisitions. It is a storeroom responsibility to pick, kit, store, stage, secure and deliver these items to the job site when the job is scheduled for execution. For direct-order items (not authorized for stocking), the planner prepares the associated Purchase Order Requests. It is a purchasing function to source, procure and expedite deliver, of these items,

❏ Determine what in-house fabrication, external contract resources, and equipment rentals are needed. Create Work Orders for in-house fabrication and rebuilds, and Purchase Order Requests with MWO reference for contractors and equipment rentals. Often these steps are not performed. As a result Equipment History can be seriously incomplete.

❏ Identify special tools and equipment required, including safety items.

❏ Consider how to get parts and people to the job location, together with ladders, scaffolding, rigging, cranes, and other heavy equipment?

❏ Coordinate related work of other groups by preparing Cross Work Orders if significant or add unique Tasks on the same WO if only minor support is needed

❏ Consider disposal issues (asbestos, oils, and other contaminants)

❏ Estimate total cost in terms of labor, material, and external charges

❏ Coordinate and expedite necessary authorizations based on final cost estimates including operational, financial, and engineering approval.

JOB PREPARATION

During Job Preparation, the Planner assembles and documents all the above planning efforts within a "Planned Job Package." Any factors that may delay or hinder effective job completion should be anticipated and steps taken for their avoidance. Given data contained within the package, coupled with a thorough verbal exchange between the planner and supervisor and a similar exchange between supervisor and mechanic, nothing should be lost between strategic preparation and tactical execution.

The Planned Job Package should include:

❑ Detailed Work Order spelling out step-by-step procedures

❑ Job Planning Sheet with Sequenced Tasks detailed by craft and skill level. Contractor as well as in-house resources must be included. Plan to perform as much pre-shutdown fabrication and other preparation as possible

❑ Duration and labor-hour Estimates for each Task. When beneficial, prepare a time-line by task to convey sequencing and simultaneous tasks to the assigned crew(s).

❑ Prepare a Bill of Materials including availability, commitment and staging location. The listing should distinguish between authorized stock items, direct purchases, and in--house fabrication.

❑ Obtain clearances and all required permits (governmental and company) completed to the point of safe feasibility. Of course, the final lock outs must be made by the responsible mechanic and equipment operator

❑ Other reference documents that the assigned crew is likely to have need for such as prints, sketches, photos, specifications, sizes, and tolerances

❑ A Site Set-Down Plan (for major tear downs)

As appropriate, the assembled package is reviewed with maintenance supervisor(s) and the requestor. The planner then holds the planned job package until all necessary materials are procured, and the job enters the coordination and scheduling phase.

FEEDBACK ON THE PLAN

One essential element of a good planning process is feedback, to facilitate improvement of planned job packages over time. Periodically a Job Planning Survey should be added to job packages. Surveys might also be sent periodically to key personnel (Maintenance Supervisors and operational requestors of maintenance support).

Survey questions might include:

1. Were job instructions clear?
2. Was the estimate within 15% (look at total hours as well as hours by craft)?
3. Was the work performed as specified?
4. Did unusual problems occur on this job?
5. Were trips for tools, parts, or supplies needed after the job was started?
6. Were there delays due to problems with permits, or per missions?
7. Were there delays relating to equipment access?
8. Were there delays caused by lack of craft coordination?
9. Allow space for explanations.

COORDINATION OF EQUIPMENT ACCESS, PERMITTING, SAFETY, AND STATUTORY PERMISSION

Safe and legal access to equipment must be addressed within the planning process. Making such provision involves thinking through the shutdown and lockout steps, people to contact, valves to isolate, access requirements, and any other items to gain and take custody of the asset.

During normal operation, the asset is in custody of Operations. Maintenance must obtain clearance to interrupt production and take control of the asset (frequently by lockout) either to rehabilitate it or to replace it. Without safe access to the asset, no work can or should be started. Access can be simple or complex. The latter may require process changes, rerouting or extensive coordination with several departments.

Reviewed below are some of the formal and informal processes by which Maintenance receives permission to work on an asset and gains control of that asset. They are driven by process, safety and regulatory issues.

Chapter 8

PROCESS DRIVEN ISSUES

In complex environments a substantial analysis is frequently required to determine when and how to take an asset out of service. It is only logical that operational or engineering permission is required to assure that chemicals won't spill out from an upstream process that wasn't shut down or that a downstream process doesn't run dry and crack a vessel that wasn't turned off.

A large production facility had three generator sets for critical loads. The controls and settings on the ASCO transfer switch were fed power from separate 24-volt batteries charged by the engines of three generator sets. The generator repair contractor didn't realize this and disconnected all the batteries at once. The entire back-up system (including the UPS bank) was taken off-line and had to be reprogrammed. Fortunately for the contractor, the power didn't go off at that time. There is a policy in place (now) that only one generator can be serviced at one time).

Another instance of bad notification with tragic consequences was the disaster in Bhopal, India. The reaction to produce methyl isocyanate was exothermic (it produced heat). If too much heat were generated, the process would get out of control. Water got into the reactor causing the process to overheat. The plant was designed with four safety systems to deal with a potential leak if it started. Over a period of years and through a series of unrelated events three of the four safety systems were taken out of service. This accident demonstrated significant deficiencies in maintenance, operations, design, and safety procedures. Reviews of the accident identified significant design defects, operational mistakes, and maintenance problems that all contributed to the tragic outcome.

For starters, the gauges measuring temperatures and pressures (including those within the crucial MIC storage tanks) were so notoriously unreliable that workers ignored early signs of trouble. The refrigeration unit for keeping MIC at safe temperatures had been shut off for some time. The gas scrubber, designed to neutralize any escaping MIC had been shut off for maintenance. The flare tower, designed to burn off MIC escaping from the scrubber, was also turned off, waiting for replacement of a corroded piece of pipe.

It must have seemed to the workers that the safety systems were not needed. Neither the crews nor supervision understood the process well enough to know that a serious problem existed.

The last line of defense was a deluge system that was designed to dump tons of cold water on the reactor to shut down the reaction. On that tragic night, when the reaction needed to be cooled down, the last remaining system between proper emergency shutdown and disaster was inoperative because it had been taken out of service. What happened is history, with hundreds of thousands of families affected by the deadly chemical.

So when the operating department gives permission to remove and replace a reactor vessel in a large, continuous process, chemical plant (or in similar complex industrial environments), tremendous thought must precede the decision. The process for granting permission must be thoroughly documented and that process must be copiously followed throughout the planning and execution stages.

SAFETY DRIVEN ISSUES

Most large facilities have processes for initiating safety reviews and applying for dangerous work permits. These permissions are closely related to legal permissions and many of them flow directly from laws. For example, proper lockout and tag out processes are prescribed by law and supported by safety permitting processes. Most large safety departments have additional requirements beyond the legal requirements. Obviously the Bhopal disaster mentioned in the previous example was also a failure of the safety permission system.

Various forms of dangerous work may require open flame permits (also called hot work permits) specifying a safety watch person with a fire extinguisher; confined space entry permits (also a US and Canada legal requirement); line-open permit where poison gases might be present; entry permit for clean or secure areas; and other forms of permit specific to an industry.

With a properly functioning safety department and a comprehensive safety permitting system in place, an incident like Bhopal is much less likely. Nothing short of certainty is acceptable.

REGULATORY DRIVEN ISSUES (LEGAL OR STATUTORY)

Statutory permission is gained from or is in compliance with laws of a legal authority. Legal issues can flow from Federal law (confined space entry rules), to State laws and right down to the local zoning board. Permit requirements from external bodies are obtained prior to a job being scheduled. Usually jobs that require legal permits have outside engineers or architects that carry responsibility for obtaining such permits. However, planning checklists should still include the permitting requirement to further assure compliance.

In a true story, bulldozers and graders were being delivered to a site to improve the flow of storm water. During heavy rain, the southwest corner of the plant would flood. A hydrologist working for a landscape architect's office had reworked the water flow into a series of attractive vegetated swales. Everyone was ready. The architect asked the contractor for a copy of the Permit from the Township. The contractor, with a sick look said, "I thought you pulled the Permit." How many times has this story been repeated? In this instance it took three months and most of the construction season to get the permit.

Chapter 8

Generally, statutory permits come in three levels.

- ❑ **Local:** zoning variances, building permits, electrical permits, plumbing permits, noise permits, etc.
- ❑ **State:** Department of environmental resources permits, and other agencies regulating individual industries such as mining.
- ❑ **Federal:** EPA permits, Coast Guard (dealing with waterways), Army Corps of Engineers, etc.

Someday there may be a fourth level for global permits designed for activities that would impact other countries. Changes in the law at any level might precipitate required changes in maintenance access procedures. Simple things such as lockouts tag-outs, and confined space entry, for instance, require steps to comply with law.

Detailed Planning Process Materials, Tools, and Equipment

The vast majority of work performed by the maintenance force requires specific spare parts and general maintenance materials. To fulfill its mission effectively, maintenance is dependent upon reliable and prompt logistical support (spares, replacement parts, materials, supplies, and special tools). The required material may be in the form of expensive parts or components unique to the particular unit being worked on, or it may be a common item of hardware costing a few cents. Regardless of the cost it is important to have the item readily available (or promptly made available) to support the efforts of the mechanics in a timely manner. Lack of such material support will create delays resulting in economic loss. Time will be lost and work quality will suffer and result in diminished equipment reliability and output capacity.

Losses range from $100 or so when labor must be shifted from one job to another, up to hundreds of thousands of dollars when a major disruption of production results. The size of the disruption is not always proportional to the cost of the required part. Such losses are clear justification for maintaining a reliable inventory of parts and materials (either on-premise or readily available from a reliable, local source) supportive of the maintenance mission.

Chapter 9

Planner/Scheduler Responsibilities to the Material Management Process

Purchasing and Stores cannot do the job alone. One Purchasing executive put this printed banner over the entrance to her area: "Your lack of planning does not necessitate an emergency for me." A reactive maintenance department is the single largest workload for purchasing departments. One reactive department encountered ran a $330,000 annual airfreight bill that was mostly unnecessary. It was due to a complete lack of planning stemming from a reactive culture rather than a proactive environment.

Purchasing, Stores, Maintenance, Engineering, and Operations should share responsibility for material support and control. Responsibilities specific to the Maintenance Planner include:

❑ Ordering special parts with adequate lead-time so that purchasing can do its job.

❑ Suggesting the inclusion of new parts in authorized stock with recommended minimum and maximum quantities.

❑ Reviewing at regular internals and recommending adjustments to inventory parameters (Min, Max, ROP reorder point)

❑ Providing adequate lead-time for Purchasing to order non-stock materials and to restock authorized stock items when there is unusual demand. .

❑ Notifying the Storeroom of delays to schedules that will delay the use of staged materials.

❑ Providing the Storeroom with adequate time to pick parts and stage them for pickup or delivery to drop points and job sites.

❑ Assuring that unused materials are returned to the Storeroom in good and clean condition for restocking.

❑ Correcting planned job packages when excessive material is returned.

❑ Notifying the Storeroom of stock items that will become obsolete or in excess, due to reduced operations, as early as possible to prevent buildup of surplus inventory. A program to identify parts that become obsolete should be an ongoing routine.

Chapter 9

**The Maintenance Storeroom can be a Productivity Trap!
Up to 25% of machine downtime hours
is due to lack of parts and materials necessary to
perform repairs correctly the first time.**

Material related steps in the planning of specific jobs are summarized in the following figure.

- ◆ Prepare the Bill of Materials
- ◆ Establish the acquisition plan
- ◆ Determine what items are in stock and reserve them
- ◆ Identify those items which must be direct ordered (sourcing, procuring and expediting these items is a Purchasing responsibility)
- ◆ Prepare acquisition documents (Stock Requisitions for items in authorized inventory, Purchase Order Requests for direct order items, and Work Orders for in-house fabrication
- ◆ Consider disposal issues
- ◆ Prepare Purchase Order with MWO reference for contractors and outside equipment rental
- ◆ List special tools and equipment required
- ◆ Arrange for ladders, scaffolding and rigging as needed

Figure 9.1 List determinable materials, parts, and special tools required

Making material available and controlling inventory levels can be conflicting objectives without a managerial policy that satisfies the overall needs of the plant and provides an established system of optimization with continual review and control. Economics should form the basis for inventory management, and each stocking decision should yield the lowest overall cost to the operation. Three major factors come into play: delivered purchase price including cost of acquisition, inventory carrying cost, and the dollar consequence of not having the item on hand when needed.

Chapter 9

Material shortages are often traceable to inadequacies within the inventory management system, an unfriendly storeroom catalog, or purchasing/stores policy driven by apparent costs (low bidder). This last item is common with functions that have inadequate concern for the impact on reliability and capacity when failing to promptly provide the needed materials to do each maintenance job promptly and properly the first time. The management and logistical teams must be persuaded that the basic prerequisite of a Just-In-Time strategy is a proactive environment within the maintenance arena. If the operation is still in the reactive mode, there had better be adequate inventory on hand. Otherwise, machines will be out of service, waiting for repair parts. The result will be a decrease in output, which will cause customers (and ultimately the firm's own bottom line) to suffer.

Effective planning and scheduling of maintenance work depends upon reliable availability of parts and materials when needed. One of the most important partnerships in the operational arena is that formed by Maintenance, Stores, Purchasing and Receiving. Without close cooperation and communication between these functions, maintenance cannot achieve functional effectiveness in support of operational reliability. When in a proactive mode, maintenance planners can provide sufficient lead-time enabling purchasing to procure in a Just-In-Time manner. On the other hand, Stores must maintain a reliable inventory to meet the demands of a reactive environment (emergency repairs made urgent by downtime). The managerial policy needed is that which best satisfies the overall needs of the facility through a system of "optimization"… with continual review and control. The four associated objectives are:

- ❑ Get the right materials to the right place, at the right time, at the right price
- ❑ Avoid excessive inventory, and
- ❑ Correctly charge all usage of parts and materials to appropriate work orders, thereby updating equipment history and charging appropriate accounts automatically
- ❑ Ease of reference by all parties via an effective, current, and complete (electronic) Stores Catalog.

Chapter 9

THE PLANNER'S ROLE IN REBUILDING

Many expensive components can be economically rebuilt. Rebuildables are a complex area for both the stockroom and for the planner. The facility might have anything from a small bench for rebuilding hand tools, valves and other small items to complete heavy-duty part rebuild capacity. Most maintenance departments have the benches and use rebuilding as a filler job for members of the responce crew between

The larger rebuild shop is usually "owned" by the maintenance department but the storeroom owns the core (item to be rebuilt, such as an engine for a generator). To complicate matters, large core units are sometimes removed and placed in the rebuild queue or sent to a vendor without the stores involvment.

These core units floating around are one of the biggest areas of confusion when an organization is trying to maintain accurate accounts for a maintenance inventory. A major issue is, how much value to place on core units before rebuilding and after rebuilding.

While strategies to account for rebuildables are beyond the scope of this work, we are concerned that rebuilders are identified on the bill of material and includes paperwork activiting the delivery of the part to the rebuilder (whether an inside or outside shop).

CONTROLLING THE MAINTENANCE STOREROOM WITH STATISTICAL INVENTORY CONTROL

The management technique by which inventories have historically been optimized is by Statistical Inventory Control, illustrated below.

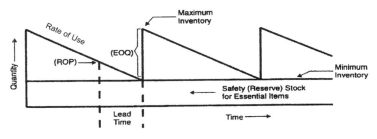

Figure 9.2   SIC Order Cycle

83

Legend:

1. Rate of Use – the average number of pieces of a specific Stock Keeping Unit (SKU) issued in a specific unit of time. The unit of time selected corresponds with that in which lead-time is stated (normally days).

2. Lead Time (LT) - the elapsed time from identifying need to order until the ordered materials are received. There is an internal as well as an external aspect to lead-time.

    A. Paper work lead time - from identification of the need to replenish through issue and delivery of the purchase order to the vendor.

    B. Vendor lead-time - from receipt of purchase order through arrival at Receiving.

    C. Stocking time – from receipt through loading onto shelves, including entry into the system so that all parties know that the item is there.

3. Reorder Point (ROP) – stock level at which a requisition is generated to replenish inventory. This parameter is calculated by multiplying the Rate of Use x Lead Time.

4. Economic Order Quantity (EOQ) - the quantity that balances the cost of carrying inventory against the cost of processing purchase orders for materials and cost associated with loss of volume discounts. With proper balance, total cost is optimized.

5. Minimum Inventory – quantity to which inventory is expected to fall immediately prior to receipt of the replenishment order.

6. Maximum Inventory - quantity where inventory on-hand including materials received is at the highest level planned (Minimum Inventory + EOQ).

7. Safety or reserve stock - quantity maintained to protect against stockouts. The causes of stockouts are variations in demand and delivery time. This provision is normally deployed only on critical items or where delivery performance tends to be unreliable.

If maximum inventory level is exceeded or a stock out occurs, a system alert should trigger reexamination of reorder point and safety stock.

Chapter 9

JIT VERSES SIC

Debates often occur between the advocates of JIT and SIC. Both are needed to optimize maintenance support of reliability objectives. Project work and planned maintenance jobs can rely on JIT because there is lead-time available. Those jobs should not be scheduled until all required materials are at hand. However, there is no lead-time for emergency breakdowns. Consequently, well managed, on-hand inventory is essential if the maintenance function is to respond promptly to emergency demands in an effective and efficient manner (making the repair properly the first time). How often have you heard the traditional cry? "We are tired of spit and bailing wire (Band-Aid) repairs. We need the job done right on the first try." Consequently, another optimization is needed; that of JIT on proactive work with SIC on reactive work.

Given the above logic, storeroom inventory is exclusively, or at least primarily, for emergency needs. That suggests that project managers/engineers and maintenance planners should not be relying on inventory to fulfill their proactive needs. Any exceptions to this logic must be taken into account when authorizing, sizing, and budgeting inventory.

Once maintenance excellence and a proactive environment have been achieved, emergency response should consume no more than 10 percent of maintenance resources. Because emergency response jobs average significantly fewer man-hours than do planned jobs and projects, they may still represent as much as 20 percent of the work orders. The material cost of emergency jobs also is typically less. Accordingly, no more than 25 percent of maintenance material needs should be dependent upon storeroom inventory ... **ultimately**.

OTHER MATERIAL MANAGEMENT CONCEPTS
TO BE DEPLOYED

In addition to Statistical Inventory Control (SIC) two other inventory control techniques are commonly deployed in well-managed storerooms to support the Maintenance/Reliability Mission. These techniques are:

❑ Classification of stock items
❑ ABC Analysis of stock items

Classification segregates stock items by criticality. Stock Outs on highly critical items have serious consequences. The risk is too great to take. Risk should be taken or minimized based on consequence of stock outs (non availability of the required part). To exercise this discretion effectively, inventory must be classified in the following manner.

❑ *Insurance spares*—high cost spares or components used on critical equipment. Not having these "insurance spares" in stock can result in extended downtime of critical equipment and major loss of production. Such parts can cost thousands of dollars. Little risk can be taken because weeks are required to obtain these items.

❑ *Insurance parts*—parts used on critical equipment or in critical components. Such parts are purchased from the OEM and are generally used on only one type of asset. . Their use is unpredictable because the mean times between failures (of items where used) is unpredictable. Costs for insurance parts range from hundreds to thousands of dollars. They normally are carried in inventory under tight control. Not having these "insurance parts" in stock can result in extended downtime and major production losses. Some risk can be taken because these items are in the distribution chain and can be obtained within a few days.

❑ *Standard replacement parts*—parts that can be used on more than one component or piece of equipment. These parts generally are stocked by suppliers for a number of users. Prompt replenishment is reliable. Same day or over-night delivery allows greater risk to be taken.

❑ *Hardware items*—bolts, nuts, washers, cotter pins and other fasteners that are low in unit cost, and readily available from suppliers. Such parts should be stocked in ample supply in a convenient location as free bin stock. Their security in the storeroom is minimal and they need not be specifically charged to the WO on which they are applied. They are "C" items as reflected in Figure 9.3.

❑ *Operating chemicals and supplies*—chemicals and supplies used in the production process.

❑ *General supplies*—office and sanitary supplies (pencils, pads, cleaners, etc.).

Chapter 9

ABC Analysis

Illustrated in Figure 9.3, is the application of Pareto's Law to the control of inventory. Pareto was a 17th century economist who proved that: "A few of the things we do are the most important, and most things we do are not critical".

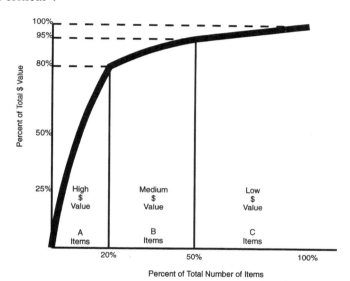

Figure 9.3   ABC Analysis Curve- Applied to Storeroom Inventory

Analyze any population, and it will be found that the most significant 20 percent represents 80 percent of the value (A items). The next 30 percent of the population represents the next 15 percent of the value (B items), and the remaining 50 percent of the population represents the remaining 5 percent of the value (C items). Management and Control effort should focus on the important items.

❑   A Items - Highest value, tightest control, close follow-up, accu rate records.

❑   B Items - Normal control, good records.

❑   C Items – Limited control, free stock.

Chapter 9

Classification of stock items and ABC analysis are distinct techniques that overlap as reflected in Figure 9.4

Figure 9.4 Stock Classification, ABC Analysis, and Essential Service Levels

Criticality	Stock Classification	Percent of Items	Value	Essential Service Level
A	Insurance Spares Insurance Parts & Other Critical Spares	20%	80%	100% 98%
B	Standard Replacement Parts	30%	15%	95%
C	Hardware Items Small Tools General M&R Supplies	50%	5%	90% 90% 85%

Before rejecting the lofty Service Levels listed above, consider the number of items "required to complete a repair right, the first time. " Should three spare parts be required and specified service level is only 90 percent, what is the likelihood of having on hand all three parts required? The answer lies in compounded probability (90% x 90% x 90% = 73%).

When considering reductions in inventory, bear in mind that insurance spares and parts represent 80 percent of inventory value and that some appropriate authority has made an informed decision that the adverse impact of not having those insurance items when needed is too great to risk.

If the overall inventory value is to be reduced by 10% (as an example) and the critical 80% of value (A) is too risky to include, the remaining 20% of value(B&C) would have to be reduced by 50% to achieve the overall objective of a 10% reduction in inventory value. Reliability and effective capacity would be seriously compromised.

Chapter 9

MATERIAL MANAGEMENT REQUIREMENTS IN SUPPORT OF PROACTIVE MAINTENANCE

As major internal customers of Purchasing and Stores, maintenance craftsmen are the primary users of supply materials and spare parts. Performance standards must define the levels of materials management support required if the reliability mission of Maintenance is to be fulfilled. These standards should be consistent with the following characteristics. Purchasing must:

❑ Process purchase order requests promptly with minimum in-house lead-time.

❑ Obtain and communicate lead-time commitments from vendors and evaluate their performance in meeting these commitments.

❑ Track materials from placement of Purchase Order Request until delivery to originator.

❑ Follow-up on availability and delivery of parts for planned work orders.

❑ Notify maintenance of delayed items prior to their scheduled delivery date.

❑ Expedite as necessary.

Storeroom must:

❑ Keep inventory orderly with parts easily identified and locatable.

❑ Keep adequate quantities of each stock item on hand to meet the day-to-day needs of Maintenance.

❑ Promptly reorder materials that are at their reorder point so stocks can be replenished before they run out.

❑ Maintain unique item identification of spare parts, materials, supplies, and tools.

❑ Provide an up-to-date catalog listing parts and supplies in stock by location.

❑ Cross-reference the catalog by Stores Reference Number, Vendor Serial Number, Key Word Description, and Where Used.

❑ Apply storeroom control in relation to item value. High value items should be controlled most closely. Medium value items should be moderately controlled, and low value items (such as hardware) warrant the least control.

- ❑ Provide quick issue service at the storeroom window.
- ❑ Process non-stock receipts and back order receipts as quickly as possible and notify users of their arrival.
- ❑ Provide timely delivery of materials to secured locations, either at the job site or at specified drop points at maintenance shops and other areas of the facility. Delivery should be synchronized with scheduled start of planned jobs and expedited to urgent jobs.

STOREROOM BENCHMARKS

➢ Service levels should be 85 to 100 percent. Depending on classification of the item requisitioned (Figure 9.4).

➢ Inventory accuracy, as procedurally determined by cycle counting, should exceed 97 percent.

PLANNING FOR SPECIALIZED EQUIPMENT AND TOOLS

The typical maintenance plan does not mention all equipment and tools, although there are some exceptions to this rule. You would not plan for tools that a tradesman could reasonably be expected to carry as part of a personal tool kit. You could assume that a pipe fitter has a 24-inch wrench but you cannot assume they carry a power threader. The planner's job is to identify any specialized, uncommon, large "shop tools" and equipment needed for the job.

> This requirement was particularly important when the maintenance department at a consumer goods manufacturer needed two high lifts in one day and had only one available. A conflict in the schedule of equipment could be trivial to solve in advance on paper, but quite expensive to solve on the shop floor when both jobs are scheduled to begin (and two crews are standing around).

Planners should also list any safety items likely to be needed, including harnesses or other PPE (personal protective equipment).

An example of such needs concerns a job to rebuild a control box in a rail tunnel. The job plan called for flashlights with fresh batteries (workers already carried flashlights). When questioned, the planner reported that the flashlights were required for safety in tunnels (any misstep could electrocute the worker on the third rail). Flashlights with fresh batteries were appropriately specified on the job plan.

Questions that planners must answer about tools and equipment are similar to the ones they must answer about spare parts.

Figure 9-4 Questions to ask about tools and equipment

1. What tools and equipment are needed for this job?
2. Do we have the necessary item in-house or do we go outside for it?
3. If we have it in-house, who is responsible for insuring it will be available (not only available but also batteries charged up, fully operational, etc)?
4. If we go outside for it, who is the preferred vendor?
5. If not currently owned, should it be purchased or rented?
6. Can we reserve it with reasonable assurance that it will be available?
7. What is the lead-time if not immediately available?
8. Is the cost included in the job estimate?
9. Do we rent it "wet or dry" (with or without an operator and insurance)?

Work Measurement

IF YOU CAN'T MEASURE IT, YOU CAN'T IMPROVE IT

Work measurement can be traced as far back as the 18th century and the work of Coulomb. Modern work measurement (time and motion studies) goes back to the late 19th and early 20th centuries with pioneering work done by Harrington Emerson, Frederick Taylor, and by Frank and Lillian Gilbreth. Some of the early efforts were applied to the work of tradesmen, such as brick laying.

Measurement of maintenance work began in earnest shortly after WWII when the US Navy developed and subsequently published "Engineered Performance Standards" for the Navy Public Works Department, which was responsible for all US Navy bases and installations. Measurement began in painting and carpentry and then expanded into other skills.

Development into a measuring "system" was a massive undertaking. At the peak of the effort there were more than 200 industrial engineers assigned to the project. Completion and initial release of data took over 10 years. In the early 1970's, NAVFAC was assigned the job of keeping the standards up to date for use by all three branches of the military (Navy, Army, and Air Force). The last major change was computerization of the entire effort in the early 1980's. The resulting library of maintenance standards is available for purchase from NAVFAC.

A related approach was applied to commercial industry during the 50's, 60's, and 70's under the common name of "Universal Maintenance Standards (UMS)." Due to the cost and labor intensity of developing and maintaining them, these standards have fallen from common usage, but an organization that has made the associated investment should not abandon the system because it did generated some of the best standards ever developed for maintenance work.

APPLICATION OF MAINTENANCE WORK MEASUREMENT

Realistic job estimates are essential to the planning, coordinating, scheduling, control, and motivation process. Before exploring the work measurement options, we should clarify the several applications of maintenance standards:

❑ Match manpower resources to workload (Macro Planning).

❑ Establish meaningful maintenance schedules ... publishing management's expectation of what is to be accomplished with the resources to be paid for during the schedule week. Job Schedules are defined in **"Duration Hours."** Duration hours times crew size equals labor-hours. These two variables must not be confused. .

❑ Forward load preventive, predictive and other programmed maintenance routines in a balanced manner to avoid peaks and valleys in the demands upon labor resources. A common mistake is to have all quarterly, semi-annuals, and annual PM tasks come due in clusters, rather than spreading them equally across weeks of the year.

❑ In the interest of customer service, promises made must be realistic, particularly when jobs require that equipment be taken out of service. Promises should include when the work will be started, how long the unit will be out of service, and when the unit will be returned to service.

❑ Realistic expectancies by which supervisory and crew performance can be evaluated and employees can be motivated. We all work more effectively when evaluated against fair, yet challenging standards. Most employees work better in an environment where they know what is expected of them and whether they are doing a good job.

The question is, which measurement approach best serves each of the several applications; true standards that indicate how long jobs should take, or estimates that quantify how long jobs will probably take. The difference is most significant when maintenance is still reactive and efficiency is seldom higher than 50%. If a standard is used in such a situation, backlog is understated, schedule compliance is extremely low, and completion promises to customers are seriously missed. The challenge is to develop a job estimate that will establish a challenging expectancy on one hand but produce an achievable schedule on the other.

Maintenance personnel are fond of stating: "Maintenance jobs never go the same way twice." While this much of the statement is basically true, they add: "Therefore, maintenance cannot be measured." The latter portion of their statement is false. Maintenance work is repetitive and therefore it can be measured. Many jobs occur weekly, some monthly or quarterly, and even annual jobs are repetitive.

The first part of the statement accurately infers that the content is not exactly the same each time the job is performed. This opinion simply influences the:

❑ Precision with which estimates can be set

❑ Methods of applying estimates to each application

❑ Time-period required to accumulate sufficient jobs into the efficiency report to level out fluctuations in individual jobs. Because of this time requirement, Maintenance Efficiency Reports normally cover a complete crew for a full week. The focus is not on job-by-job efficiency of individuals.

Industrial engineers strive to achieve accuracy plus or minus five percent (±5%) for production incentive standards, but in maintenance we strive for plus or minus fifteen percent (±15%). This level of accuracy is adequate for the applications to which we apply estimates in the maintenance arena. Accuracy for individual jobs may not be reliable, but combined accuracy over the dozens of jobs most maintenance crews complete within a week is adequate. Especially when the focus is on trends over time, rather than excessive emphasis on results within a specific time period.

Maintenance work often does contain unpredictable elements, but those elements seldom influence the entire job. If disassembly should double due to working conditions; troubleshooting, re-assembly, clean up, and feedback are not impacted proportionately. The total job, most likely, will not increase beyond the fifteen-percent range (+/-). The objective is to consistently achieve reasonable accuracy of time needed.

An area where estimates can go wrong is where the scope of work changes dramatically. If the job is to change a simple bearing and ends up replacing all bearings plus the shaft, impeller, and housing, the scope change dictates a new estimate. Judgment must be used. Small scope changes can be absorbed in the +/-15% expectation but large changes in scope require new estimates.

OPTIONAL LEVELS OF WORK MEASUREMENT METHODOLOGY

As exhibited in Figure 10.1, several forms of work measurement exist (together or separately) for use in maintenance applications. Having discussed why and how to use labor standards, the second phase of our discussion focuses on what method of standard/estimate development is most suitable and economically viable for maintenance needs. Clearly, a reasonable return on the associated investment must be considered in the selection process.

There are several viable approaches to the measurement of maintenance jobs and to means of maintaining comparative consistency between estimates and estimators. The level of work measurement sophistication (summarized in Fig. 10.1 and characterized as progress up the staircase) commonly increases with: the maturity of the maintenance management installation, the level of precision required to raise maintenance performance to the next plateau, and the focus of the installation (Efficiency Measurement or Schedule Compliance). An Efficiency focus (Standard Labor-hours ÷ Actual Labor-hours = % Efficiency) requires a more accurate form of measurement. A Planning and Scheduling focus requires a less precise form.

Chapter 10

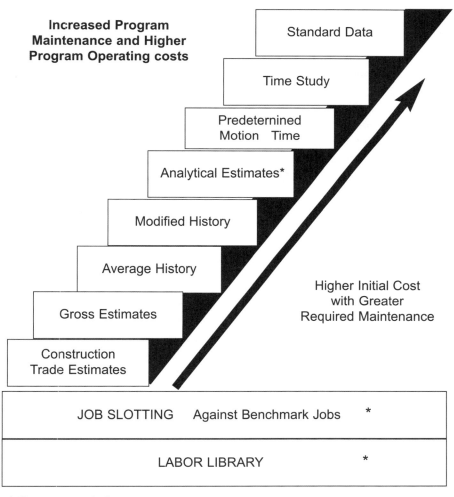

Figure 10.1   Levels of Maintenance Work Measurement

97

Chapter 10

Advantages and Disadvantages of each Methodology

❑ **Construction Trade Estimates (Published Standards)**

Construction Trade Estimates are shown as the lowest step because they are developed for contractors to use when bidding on construction jobs. They are not recommended for application to in-house maintenance. The reasons are that construction is not among our most efficient industries and the estimates reflect that level of efficiency, they also include engineering safety factors in the interest of the bidder, and most relate to construction jobs rather than maintenance jobs. When seeking such standards, first consult the manufacturers of the equipment or components (seals, bearings, etc.).

❑ **Gross Estimates**

Although they may be only educated guesses, most installations begin with Gross Estimates ... the second step on the desirability staircase. These estimates are the least costly and least time consuming. The disadvantage is that they usually reflect personal judgment of a particular supervisor, planner, or technician, and different people guess differently. Consistency seldom exists between estimators and their estimates typically reflect how much time jobs are taking, rather than how much time they should take if they were properly prepared. Some estimators base their judgment on how long they themselves took to perform the job when they were craftsmen. Others base their estimates on how long a specific technician usually takes to perform the job. These estimates are especially poor when the estimator has not personally experienced the job.

However, if backlog is currently quantified in terms of "jobs" or work orders, get it quantified in labor-hours...now! Even if use of gross estimates is temporarily necessary. "Jobs" is not a meaningful unit of measure because some require less than an hour, while others may require a hundred hours or more. The only common denominator is "labor-hours."

Chapter 10

❑ Historical Averages

Records of past actual man-hours charged (Historical Times) are the third step in sophistication and the second most common methodology. The labor-hours charged to previous work or individual jobs are recorded and accumulated. They are then averaged after elimination of skewed highs and lows. Some computer systems encompass this functionality. Resultant averages obviously reflect size and condition of the facility, condition of the equipment, skill level of the maintenance work force, and the current state of job preparation and materials support. All of which are likely poor, if the local culture is reactive.

If the average length of time that a job has been taking is known, what could possibly be a better estimate? The answer is nothing … provided the maintenance crews are working at 100 percent efficiency and the state of maintenance excellence has already been achieved. Unfortunately, the efficiency of reactive maintenance tends to be in the 50% range. If world-class is to be achieved, such mediocrity cannot be the expectancy.

Efficiency is standard labor-hours divided by actual labor-hours. If historical averages are used, the calculation effectively becomes (Historical Average divided by Actual). Efficiency always approximates 100% … but undeservedly. Efficiency and schedule compliance may look favorable using historical averages, but management will soon ask, "How demanding are the expectancies?"

Because the work order system is the source of historical averages, it is often difficult to obtain reliable data on which to calculate the averages. If descriptions are not correct and time distribution is not complete and accurate, the averages will not be reliable. They will reflect the current environment, methods, and tooling rather than standard methods and procedures with proper preparation through planning, coordination, and scheduling.

Chapter 10

❑ **Adjusted Averages**

Adjusted averages are the first reflection of a true expectancy (a standard). They require a base period of perhaps six months during which time averages are collected for repetitive maintenance jobs and activity sampling is performed concurrently (see page 2). Sampling establishes the average efficiency of each crew. If a given job averaged ten labor-hours and the crew averaged 70% efficiency during the base period, 7.0 labor-hours becomes the adjusted average.

This methodology provides quasi-standards but are fairly expensive due to the cost of sampling, and require several months before measures are available for deployment.

❑ **Analytical Estimates**

Analytical estimates are recommended as the appropriate level of work measurement sophistication. They will be fully discussed in the next chapter.

❑ **Job Slotting and Labor Libraries**

Job slotting and labor libraries are effective work measurement techniques recommended for use in combination with Analytical Estimates. They also will be fully addressed in the next chapter.

❑ **Universal Maintenance Standards**

Predetermined Motion Times, Time Study, and Standard Data take work measurement too far for the needs of Maintenance Management. In combination they evolve into Universal Maintenance Standards (UMS) and these have faded from common usage. Although they are the most accurate method by which to develop maintenance standards, such standards are too time consuming and expensive to set-up as well as maintain. Consequently they are not generally recommended. However, development of these standards did spawn the three techniques that are recommended (Analytical Estimates, Job Slotting, and Labor Libraries).

Regardless of methodology used, to insure consistency of development, estimating formats specific to the level of estimating in use should be established. Figure 11.10 in Chapter 11 contains an example specific to analytical estimates.

WHICH METHODOLOGY BEST SERVES THE SEVERAL APPLICATIONS OF WORK MEASUREMENT?

The key to effective work measurement is to establish standards and measure current efficiency relative to standard. With knowledge of these two metrics, expectancy can be adjusted to serve the application. Although it is desirable to use management tools with "pull" to drive improvement, it is not practical to expect efficiency to improve from 50% to 100% (or even 85%) in near-term. In fact, 10% improvement in a quarter is a lofty goal.

❑ **Backlog** Weeks should be calculated on current efficiency.

❑ **Schedule**s People work at a pace that is based partially on the amount of work they are given. Ideally, jobs are quantified within the schedule based on standard labor-hours (expectancy of how long a job should take).

However, there is also a customer service aspect to the scheduling process. Because the schedule represents promises or commitments to the internal customer there is a need to accurately predict when a unit will be removed from and returned to service. Therefore, it is crucial that the schedule be practicable, given current efficiency levels. Consequently, schedules are often developed based upon a 10% improvement (pull) over current efficiency, and this approach makes them achievable without accepting the status quo.

❑ **Efficiency** should be based upon a true standard (expectancy of how long a job should take), and should never be used for discipline (as a crutch to cover-up for a lack of proper supervision and leadership). Work measurement must be protected

101

for the important applications of backlog control, work scheduling, and efficiency trends.

By reflecting all time lost throughout the day regardless of nature or source, efficiency measures the entire maintenance delivery system … not simply the performance of individual mechanics. The organization's ability or inability to assign jobs, communicate critical information, provide parts and materials to perform the job properly (the first time), and provide safe access to the affected equipment also affect efficiency.

Efficiency Reports should be developed weekly for the total crew or team responsible to a given supervisor. Using a one-week time frame means that the average efficiency for each crew is based on several dozen jobs, thereby balancing abnormally difficult jobs with unusually trouble free jobs.

Individual (as opposed to crew) efficiency should be the exception rather than the rule, and should only be used in a constructive manner to guide individual development and training. Analyzing the efficiency of individuals over time often uncovers specific training needs.

Properly utilized, the measurement of individual and team efficiency is an important motivator. Without it, workers do not get sufficient and accurate feedback about their job performance. The feedback a worker does receive relates more to moods of supervisors and customers than to job quality or duration. Standards help by providing the expectations of management. When mechanics meet the standards, they know they are meeting management's expectations. This form of communication and feedback improves morale.

BUILDING AN ESTIMATE

There is no substitute for task knowledge when it comes to estimating. Regardless of the measurement technique, familiarity with operating equipment and the jobs required to maintain them is essential. Planners bring some of the necessary familiarity to the job, while the balance is acquired through formal and on-the-job training. Craft background is the ideal starting point because it enables a Planner to "visualize" how the job should be performed. If the planner does not know the job well enough to visualize it, he or she should talk through" the job with a maintenance craftsman or supervisor who is familiar with it.

At times planners will observe a job as it is being performed. Although this approach will not contribute to effective job performance within this occurrence of the job, it will facilitate development of a good planned job package for future occurrences. Such observations expand planner knowledge of plant processes and equipment, and inherent skills of individual technicians comprising the teams the planner supports. Tremendous advantages accrue when planners are seen to be involved in proper preparations for effective job performance. A good planned job package builds credibility and appreciation between planners, supervisors, technicians, managers and internal customers.

Nine elements of planning (description, scope, job steps, parts, tools, access, information, permission, skill sets) come into play within the estimating process. If not properly addressed within the planning process and provided for within the associated estimates, a defect in any of the nine elements can sink a job and cause it to run over estimate and into overtime. In which case, efficiency, schedule compliance, and customer satisfaction will all fall short of expectations.

The phrase "fix pump" is not adequate information on which to establish an appropriate standard. The description must state for example, whether the pump requires a simple adjustment to stop a leak, replacement of a check valve, or possibly replacement of the entire unit? The requestor may not know the answer when originating the work order, but the planner must know the answer before estimating the job. All plannable jobs require an estimate to be pre-posted. Estimates for emergency jobs are usually post-applied based upon feedback following job completion. To receive appropriate credit within Efficiency Reports, when work actually performed differs from requested or planned work the complications or extra work encountered must be recorded by the mechanic so the planner can post-apply an adjusted estimate prior to job close out.

JOB CREEP

This is what happens when the scope of work changes as the job progresses. Often the time increase is not incurred on the planned job but on other jobs the customer may request while the technician is in the area. On the surface such requests seem reasonable enough, but the customer normally has no concept of other commitments that depend upon the scheduled capacity the given technician represents. Planners need to clarify why jobs "presumably" exceeded the estimate before making any changes to the expectation.

Initial effort within the planning and estimating process can provide significant rewards. Begin by reviewing the job to assure inclusion of all work within job scope. The single biggest problem in estimating is not pinning down the start and finish points of the job. Given proper scope, there are several essentials that make estimating easier, more accurate, and more consistent between those who estimate - planners. These include:

- ❑ Breaking large jobs into steps. Long, complex jobs cannot be estimated accurately as a whole.
- ❑ Not trying to estimate with "pinpoint" accuracy
- ❑ Relative comparison of new jobs to common known jobs – "benchmarks". The new job need not "match" the benchmark. It is necessary simply to determine which benchmark is the closest comparison.

A properly developed standard is based upon an experienced, properly trained technician working deliberately without undue exertion for the period of time between authorized breaks (approximately two hours). Experience, age, gender, and health of individual technicians are not reflected in proper standards. To begin with, planners should be promptly quantifying the backlog of plannable jobs. This may be weeks before the job will be scheduled. Therefore at time of estimating, they have limited knowledge as to whom the job will eventually be assigned.

Any study of standards (job time requirements) is incomplete without discussing job quality. Standards must include sufficient time for the worker to perform the proper job in a quality manner (define the right job, perform it right, the first time). Supervisors must make sure that timely job completion is not achieved at the sacrifice of job quality.

Analytical Estimating

A nalyzing and estimating maintenance work seems hard at first because there are nearly always elements that are unpredictable. Normally however, these elements do not constitute the whole job and quite often are only a minor portion. The purpose of analytical estimating is to quickly develop reasonably accurate and consistent time estimates. The technique is simple and based on the following principles:

❑ For persons who have had practical experience performing maintenance jobs, it is relatively easy to visualize and establish a time requirement for simple, short duration jobs. Because of their experience, ex-craftsmen usually make the best planners.

❑ Long, complex jobs cannot be estimated as a whole. Estimation of such jobs is easier and more accurate when the job is broken down into separate steps or tasks and estimated at that level, then summarized into an estimate for the total job.

❑ Pinpoint accuracy in estimating is not justified or achievable because all the variables in maintenance work cannot be known until *after* the job is completed. In maintenance we therefore look for ±15% accuracy.

Chapter 11

All maintenance jobs can be broken down into the sequence shown in Figure 11.1.

Figure 11.1 The Common Job Sequence

Source	Task
P	Get ready and receive instructions for doing the job. Preparation includes receiving job instructions from the supervisor, collecting personal tools, obtaining parts from storeroom, and gathering required special tools and equipment
T	Travel to job site (outbound)
P	**Listen** to production input regarding symptoms
P	Make preliminary diagnosis and troubleshoot prior to shutdown
P	Shutdown and Lockout. This procedure is done jointly with the line supervisor, control room technician, and/or line operator. Equipment must be stopped in the proper sequence with proper lockouts
L	Partial or total disassembly to reach the problem area
P	Determination of full extent of problem
P	**Identification** of necessary replacement parts, obtaining them from the storeroom, or initiating a direct purchase for parts not held in inventory
L	**Re**-assembly of equipment using the replacement parts as needed
P	**Check** proper job completion, test operability of equipment, clean up the job site and put away tools
T	**Travel** back to shop (inbound)
P	**Report** on job and return unused parts, special tools and equipment
A	**Allowances**

Legend - Source of Estimate

	P	Fixed Provision Table
	T	Travel Time Table
	L	Labor Library
	A	Allowance Table

Chapter 11

T - Travel Time Table

Within a short period of time, most maintenance departments will perform thousands of jobs. How many times should travel time be estimated? It might be wise to devote a little up front effort to develop a travel-time table (not unlike a point-to-point distance chart on a road map). Using plant drawings, measure distances from the maintenance shop to the center of each Production Department or Work Center... as size may dictate. Convert that information to round-trip travel times using a normal walking pace (250 feet per minute).

Develop the table further, and apply the data to simple, average and complex jobs. Simple jobs can be done with the tools technicians carry in their tool pouch; a single round-trip is adequate. Average jobs often require an extra trip back to the shop or perhaps to the storeroom; allow two round-trips. Complex jobs may require a third round trip. No job properly planned should need more than three round-trips, unless it requires more than one shift to perform. A simple travel-time table is shown in Figure 11.2

Figure 11.2   Travel-Time Table
(This table would be filled in based on the plant layout and distances)

From Shop To	Round Trip Hours	Allowed Hours Per Person		
		Simple	Average	Complex
Area A	0.5	0.5	1.0	1.5
Area B	0.4	0.4	0.8	1.2
Area C	0.3	0.3	0.6	0.9
Area D	0.2	0.2	0.4	0.6
All Other Areas	0.1	0.1	0.2	0.3
	Round Trips Provided	1	2	3

P - Miscellaneous Provision Table

All the steps coded "P" are assembled within the miscellaneous provision table. As with the Travel Time table, how many times must these common provisions be estimated? Again, some up-front effort needs to be invested to estimate these needs well and assemble them into a table to be used by all Planners. The time provisions shown in Figure 11.3 are for illustration only. Appropriate estimates must be developed for the plant involved.

107

Figure 11.3 Miscellaneous Provision Table
(To be adapted to plant operating conditions)

	Simple	Average	Complex
Feedback (paperwork)	WO and TDC	Plus PM Checklist	Detailed Feedback
Receive Instructions	Up to 3 minutes	Approx. 5 minutes	Approx. 10 min.
Gather Tools	Tool Belt	From personal box	Special Shop Tools
Follow Safety Procedures	Normal	Lockout	Rope area, special clothing
Obtain Parts & Materials	None	Bin Stock	Storeroom Requisition
Reference Job Plan, Drawings, etc.	None	Job Plan	Multiple References
Allowed Hours per Person*	**0.1**	**0.3**	**0.5**

* All crafts except Painters. Add additional time as appropriate for mixing, taping, drop cloths, etc.

L – THE LABOR LIBRARY

Table 11.4 is developed for each piece of equipment and/or nature of work. It addresses the pure work that takes place at the job site (wrench time) … broken down by job step or task.

Some of the information for this table is obtained from technical manuals for the various equipment units. Most manuals contain disassembly schematics that detail the disassembly sequence (what comes off first, through what comes off last). Re-assembly usually is simply the reverse of disassembly.

Determine and document the required skills, crew size, duration, and labor-hours required for each step. It then becomes a relatively simple issue of determining the depth of disassembly necessary to reach the problem area. The resultant library should take the following form:

Figure 11.4   Labor Library
(Develop this file for each major repair on each piece of equipment)

LABOR LIBRARY										
Equipment Name:					**Equipment No:**					
Sequential Tasks	**Craft:**	Mechanic		E & I		Construction		Labor		
No.	**Description**		Crew	MH	Crew	MH	Crew	MH	Crew	MH
0.10										
0.20										
0.30										
0.40										
0.50										
0.60										
0.70										
0.80										
0.90										
1.00										
1.10										
1.20										
1.30										
1.40										

Note: Capture the disassemble/reassemble sequence. Whatever comes off first usually goes back on last. See Appendix for an example filled in.

Initially, it probably seems like an overwhelming task to develop such data for each equipment unit in the operation. However, if all the work is documented and cataloged, complete libraries will soon be built up. To complete the first job that arises on a given unit, deep disassembly may not be needed. Catalog the steps needed. The next time a failure occurs on the unit, you may need to go deeper. Add the additional depth to the library. It won't be long before the entire disassemble/re-assemble sequence is captured.

All maintenance departments do (at least occasionally) define job steps, perform material take-offs, locate and reference prints and drawings, prepare clarifying sketches, and make labor-hour and job duration estimates. The historical problem has been that all these preparatory efforts expended through the years have-not been retained. The planning information is commonly discarded upon job completion. Advantage should be

109

taken of the repetitive nature of maintenance. Catalog planning efforts in a retrievable format. They will be applicable time and again. This is the way to get your head above water. If you continually build these planner reference files, planning efforts become better, easier and faster over time.

ALLOWANCES

Direct work standards (labor library, travel time table, miscellaneous provision table) do not include provision for activities such as authorized breaks and wash up, fatigue, unavoidable delays, and crew balancing on multi-person and multi-crew jobs. Such activities are inherent in maintenance work and must be provided for by addition of allowance factors. The first three examples are commonly referred to as Personal, Fatigue, and Delay allowances (PF&D); at a rate of 15% (5% + 5% + 5%). The latter two items cover specific situations that are common in maintenance work. More than one person or skill is required during the job, sometimes simultaneously, other times not, and it is impossible to perfectly coordinate them. It is inevitable that one individual or group is forced to wait on another to arrive or simply to get out of the way periodically during such jobs. When the job is well planned, these instances are minimized. But they still occur. A typical table of allowances is shown in Figure 11.5

Figure 11.5 Table of Allowances
(Illustrative)

Nature of Allowance	%	Level of Inherent Fatigue		
		Light	Average	Heavy
Personal Time (Breaks and Clean Up)	5	5	5	5
Fatigue	5–10	5	7	10
Unavoidable Delay	5	5	5	5
Sub Total	15-20	15%	17%	20%
Crew Balance – Multi Person	0-3	0-3	0-3	0-3
Crew Balance – Multi Craft	0-2	0-2	0-2	0-2
Total	15-25	15-20	17-22	20-25

Notes: Use 15% for preparation and travel time. Use appropriate fatigue for direct work. Crew Balance also applies only to direct work

The vast majority of maintenance jobs need only the 15% allowance. A good number require 17 or 18%. Some jobs take 20%. Very rarely, a job may require up to 25% because all the above conditions seldom come together in a single job or throughout an entire job.

Allowances are typically added to the total accumulated labor-hours (direct work, miscellaneous provisions, and travel time). However, there are times when it is imperative to add them at the level of individual tasks. Firstly, it is unlikely that heavy fatigue or crew balancing is necessary throughout an entire job. Secondly, when providing a job time-line add allowance factors to duration hours as well as labor-hours to clearly show the impact of allowances throughout the course of the job. Clearly, nobody waits to the end of a large job to exercise all the allowances.

COMPARATIVE TIME ESTIMATING
(MAKING SURE EVERYBODY COMPARES THEIR ESTIMATES TO THE SAME STANDARD)

Absolute accuracy is neither possible nor necessary to establish a proper expectancy for control of maintenance. What is essential for control is *consistency*.

It is in the nature of human mental processes to estimate by comparison. We compare the unknown with the known and then estimate the degree of similarity (identical, larger, smaller, by how much?). We tend to do these comparisons automatically and very subjectively. Some people have an almost infallible sense of comparative magnitude, others find it hard to come anywhere near accuracy.

The basis for consistent estimating is for all planners to refer to and apply a common library or catalog of job estimates. There are basically four means of structuring comparative maintenance job estimate files:

❑ Systematic card files, by: Skill, Nature of work, Crew size, Duration, Labor-hours. Of course, use of spreadsheets or database software simplifies the job.
❑ Catalog of standard data for building job estimates
❑ Labor Library
❑ Slotting Table of Benchmark Jobs

Best practice rests between the latter two structures. Slotting tables are discussed next.

Regardless of the comparative approach used, the basic form of work measurement must still be selected (Gross Estimates, Analytical Estimating, etc.). How are the standards (or estimates) that are to be loaded into the comparative catalog (regardless of which of the four above approaches is selected) going to be developed?

SLOTTING TABLES

Slotting compares jobs to be evaluated with a group of jobs that are well known, and have been carefully described and estimated. Usually it is easier to determine whether a job is bigger or smaller then another, then to determine, in isolation, how long the job will take.

To make a comparative estimate, a planner must first define the scope of the job to be performed and must prescribe the method to be used. This work requires good knowledge of processes and equipment, visits to job sites, and discussion with requestors, responsible maintenance supervisors and knowledgeable technicians.

Back at his workstation, the planner's next step is to consult the appropriate section of the benchmark library to locate a benchmark job of similar nature and level of labor intensity through a bracketing process. He makes a judgment based on his own mental comparison of what is involved in doing a benchmark job (for which required labor-hours are established) to what will probably be required to do the job being estimated. Using the process illustrated in Figure 11.7 he makes a judgment as to which slot the new job best fits.

Judgment may indicate that the new job is less labor intensive than the benchmark jobs in slot **"K"** but more labor intensive than those in slot **"I."** Although most likely not an exact match with any of the **"J"** benchmarks, nevertheless this comparison is the best available. Close enough! It is another **"J"** job, and these are known to range between 5.5 and 6.5 and average 6.0 labor-hours.

Notice that the increments in Figure 11.6 are consistent with a desired accuracy of $^+/_-15\%$ and remember that the technique of comparative estimating involves the comparison of jobs with those in the library, not the *matching of jobs.* This distinction is important because a few hundred carefully selected benchmark jobs in the library will enable an experienced Planner to produce consistent estimates for most maintenance

Chapter 11

The Slotting Concept

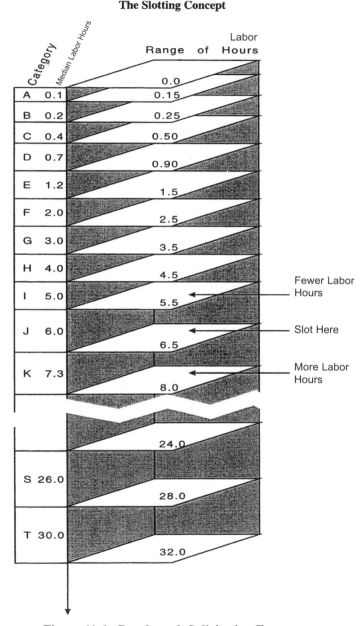

Figure 11.6 Benchmark Solicitation Form

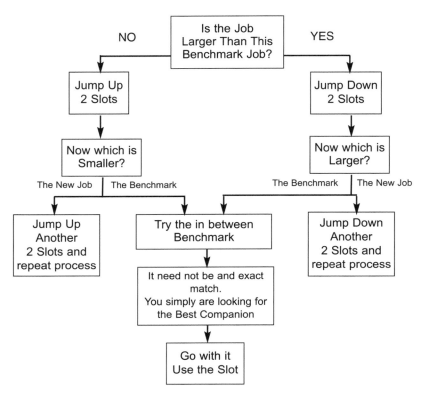

Figure 11.7 Slotting Table Concept

work. Many thousands of jobs would have to be described, planned and estimated to provide exact matches for all specific jobs performed by a maintenance organization.

Comparative estimating still requires subjective thought by the planner, but the only decision he must make is a choice between one time interval and the next. Comparison with library jobs of known duration greatly reduces guesswork.

Figure 11.6 presents the slotting concept visually, but actual Slotting Tables commonly take the form illustrated in Figure 11.8, which contains a spreadsheet of Benchmarks cataloged by required skills within the nature of the task.

Chapter 11

Figure 11.8   Slotting Table - Common Format

SPREADSHEET

GROUP _____   GROUP CODE ___01-04_____

E. Jobs Code _984_	F. Jobs Code _985_	G. Jobs Code _986_	H. Jobs Code _987_
2.5 HRS.	3.5 HRS.	4.5 HRS.	6.0 HRS.
2.00 HRS.	3.00 HRS.	4.00 HRS.	6.00 HRS. 7.00 HRS.
Repair Door	Repair Calendar Drive	Repair Top Press Roll	Replace Wire on P.M.
Repair Valve - Up to 4"	Misc. Minor Repairs - Pulper	Replace Top Wet Press Roll	Replace Pump
Repair Hoist - 1 to 5 Ton	Replace PIV Valve	Replace Doctor on P.M.	Replace Top Size Press Roll
Repair Service Elevator	Replace Doctor Blade Holder	Repair Fan	
Repair Pump	Install Dandy Roll		
Misc. Minor Repairs - Langston Winder	Minor Repairs to Cylinders		
Repair Print-Weigh Scales	Repair Saveall		

DEVELOPMENT OF SLOTTING TABLES

Maintaining the reference library is a central function requiring the assembly of contributions from all planners to cover the various classes of equipment. Maintaining the library requires that a uniform structure for job estimates be established and controlled.

Initial Slotting Tables can be developed in the following manner. Solicit up-front input from the organization (perhaps involving all the planners and supervisors as well as the technicians). Explain job slotting and its usage. Then enlist participation with the following technique:

INSTRUCTIONS FOR RECOMMENDING BENCHMARKS

❏   Identify some jobs performed by your own crew that occur frequently and are familiar to you.

❏   Determine the most effective crew size

❏   Record each selected job in the appropriate interval on the

115

Benchmark Solicitation Form (Figure 11.9). Put the job in the appropriate crew size column and in the increment row representing your best judgment. Work in pencil, as you may wish to change an estimate based on the relativity of one job to another. Use of this relativity is the concept behind the slotting techniques.

❑ Offer a recommended benchmark job in as many slots as possible, but do not force a job into a wrong slot. It is better not to use a job or to leave an increment blank, than to force a job into an improper increment.

❑ Submissions will be compared to those from other partici pants and as necessary will be spot-checked by work measurement. Therefore, do not hesitate to offer your best insights.

This approach provides a quick start-up with reasonable initial accuracy and consistency. Resultant benchmarks are subsequently refined as the planning and estimating process matures. The refinement process incorporates the following technique:

❑ Each week an exception report is generated via the CMMS. For each planner, it lists the top ten jobs for which actual labor-hours differed from his estimated labor-hours by the largest percentage, regardless of sign (plus or minus).

❑ The involved planner should evaluate those ten jobs to determine if adjustment of the estimate is necessary. If the planner decides that he/she was wrong, the estimate should be changed in the database for future application. On the other hand, if it is decided that the source of variance was poor performance by crew or supervisor, or some avoidable delay that should not repeat in the future, then the estimate should not be revised.

Using Foreman Estimates
To Building the Slotting Library

Craft:

Maintenance Management Program
Benchmark Development Sheet

	Elapsed Time			Suggested Benchmark Job			
Slot	Allowed Hours	Range From-To	Units	One Person Crew	Two Person Crew	Three Person Crew	Four Person Crew
A	0.30	0 to 20	Minute				
B	0.50	20 to 40	Minute				
C	0.00	41 to 50	Minute				
D	1.00	51 to 1-1/4*	Min-Hr				
E	1.50	1-1/4 to 1-3/4*	Hours				
F	2.00	1-3/4 to 2-1/2*	Hours				
G	3.00	2-1/2 to 3-1/2*	Hours				
H	4.00	3-1/2 to 5*	Hours				
I	6.00	5 to 7*	Hours				
J	8.00	7 to 9*	Hours				
K	10.00	9 to 11*	Hours				
L	12.00	11 to 14*	Hours				
M	16.00	1-3/4 to 2-1/4*	Shifts				
N	20.00	2-1/4 to 2-3/4*	Shifts				
O	24.00	2-3/4 to 3-1/2*	Shifts				
P	32.00	3-1/2 to 4-1/2*	Shifts				
Q	40.00	4-1/2 to 5-1/2*	Shifts				

*Not Including; - Use the Higher Slot

Figure 11.9   Benchmark Solicitation Form

117

Chapter 11

SLOTTING TABLE CATALOGING

As the list of estimated jobs expands, slotting tables become more specific. The objective is to develop families of Slotting Tables. Use of this technique means that estimates will improve progressively as the planning process matures. Figure 11.10 outlines a family of slotting tables for Millwright work.

Millwright Task Areas

		Code 0290XX
Code 0290.XX	**Task Area Title**	**Coverage (includes replace/repair/clean the following:)**
0.01	Air Compressors, Vacuum Pumps	Reciprocating and rotary type vacuum pumps
0.02	Air Cylinders	Single action, double action
0.03	Bearings, Bushings	Anti-friction bearings, plain bushings
0.04	Chain, Belts, Cables	Roller, silent, and conveyor chain; V-type, fiat-type, Veelos belts
0.05	Bakes, Cluches	Manual electric, pneumatic
0.06	hives, Gear Reducers	Variable speed drives (Reeves & PIV); fixed ratio speed; shaft mounted reducers
0.07	Fans	Materials handling and air handling
0.08	Fasterners	Keys bolts, screws, taper & steer pins
0.09	Filters	Air, oil, and water
0.10	Fire Protection	Inspect, test, and recharge
0.11	Gears, Pulleys, Sprockets	Gears: spur, bevel, Helical, worm, Pulleys: V-type, fiat conveyor type
0.12	Motors	AC motors: single- and 3-phase; DC motors; hydraulic motors; gear-head motors
0.13	humps	Centrifugal, positive displacement, gear, reciprocating
0.14	Rolls, Rushes, Shafts	Rolls - Covers all types: s.s., rubber, felt, cork; shafts used as rolls; elliptical, etc.
0.19	PM Inspections	Covers PM inspections on all equipment ad machines

Figure 11.10   Family of Slotting Tables For Millwright Work

118

Chapter 11

JOB ESTIMATING WORKSHEET

One of the most important considerations in the preparatory process is consistency of job estimates. Without consistency, the labor force will feel that estimates are unfair and will fight their usage. The following worksheet will help the planner maintain a consistent approach.

Maintenance Job Plan - Job Estimating Work Sheet

Planner:	Request #: WO #:	Date: Date:

Equipment Number Performed Upon: _____

Description: _____
_____
_____
_____

Work Group: _____
Characteristics: _____ _____ _____
            Simple   Average   Complex

Crew Size: _____

Crew Balancing Required_____ _____
            Yes   No

Travel :
Fatigue Factor: _____ _____ _____
         Light   Average Heavy

Skill Required: _____

Multi-Craft Allowance Required: ___ ___
                 Yes   No

Build Up of Detailed Estimates
Required

Step	Activity	Skill	Crew Size	Time Estimate
1.				
2.				
3.				
4.				
5.				
6.				_____

Total Direct Work (Elapsed Time):
(Carry to below summary)

Normal Time x Allowance Factor = Estimated Time

Direct Work Time: _____
Job Preparation & Wrap Up: _____
Travel Time: _____
Total Normal Time = _____ x Allowance Factor of _____ = Estimated Duration _____
                                      x Crew Size _____
                                      = Total Estimated Manhours _____

Estimate Developed By: Slotting ☐   Detailed Estimate ☐   Historical Average ☐   Gross Estimate ☐

Coordination with Operations

In maintenance, the coordination process is a cooperative effort between Maintenance, Operations and other involved functions. It is designed to achieve consensus between equipment custodians, maintenance, engineering, and procurement. The objective is to agree on the most important maintenance jobs to be completed with the available resources during the coming week. For those jobs requiring equipment to be out of service, times must be found and agreed to when Operations can release the asset and maintenance can make the necessary resources available. The coordination meeting is held weekly, normally late (i.e., Wednesday PM or Thursday AM) of the week prior to the week to be scheduled.

Maintenance work should be scheduled to optimize the balance between minimal adverse impact on the operating schedule and effective utilization of maintenance resources. This optimization requires communication, collaboration and coordination with Production-Planning to link the maintenance schedule with the operating schedule. This synchronization is illustrated in figure 12.1.

In a production environment where uptime is essential, coordination maximizes the use of "windows of opportunity" to accomplish work whenever and wherever equipment is not in use. Where opportune windows do not present themselves, downtime for planned maintenance must be scheduled to minimize disruption of production. The situation is not

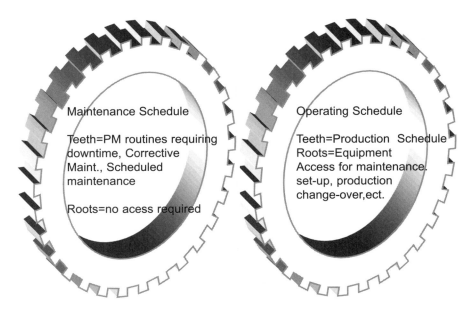

Figure 12.1 Synchronization of production and maintenance.
The production schedule and the maintenance routines are generated so that they match

much different in Building Maintenance (such as in a hospital). Servicing of some systems, areas, or rooms has to wait until it is convenient for the medical staff or patient (who might be in the ICU or might be undergoing an operation).

Planners should view liaison with operations as a permanent relationship. They should learn and take interest in the problems of their "internal customers," remaining cognizant of operating workloads, short and long-term plans, and priorities. By these means, planners can help operations to think far enough in advance to facilitate effective planning and scheduling of maintenance resources in optimal support of operating plans.

In selecting jobs for the Weekly Master Schedule, it must first be ensured that all preventive and predictive routines are scheduled at their predetermined frequencies. The PM's of all crafts (electrical as well as mechanical) must be considered, together with approaching PM's; perhaps they should be performed early to take advantage of scheduled downtime agreements to avoid another shutdown in only a matter of a few weeks.

Next, the various parties should be aware of all jobs approaching requested completion dates. Any jobs that cannot be scheduled to meet those dates should be discussed with the Requestor in the context of other job priorities established by all attendees. Conflicts must be considered from all perspectives. That of the internal customer must ultimately prevail in conflicts involving maintenance and operations. Conflicts between internal customers must be resolved by their common manager in the optimal interest of the overall operation.

Planners return to their workstations to layout detailed schedules for next week that reflect agreements reached.

- ❏ Operations agree to make the equipment available in the agreed state, so that work can be performed as scheduled.
- ❏ Maintenance agrees to perform the work as scheduled, starting on time and finishing on time. Reliable estimates of outage duration must therefore be made.

Compliance with schedule requires that resources designated for given jobs not be diverted to unscheduled work except for true emergencies. The ultimate Operations approver of the schedule is the logical authority regarding any urgency constituting a "schedule-break".

PLANNER PREPARATION FOR THE
WEEKLY COORDINATION MEETING

- ❏ **Upkeep of Backlog.** Planners keep backlog cleansed, current and accurate.
- ❏ **Issuance of Backlog Reports.** Planners issue current, well-organized Backlog Reports by Status to all attendees (operating, maintenance and engineering supervision and management) on the day before the meeting. Ideally, distribution is electronic.
- ❏ **Resolution of Conflicts.** The planner sorts through the mass of (frequently) conflicting demands of the many internal customers of Maintenance in order to better chair the coordination meeting
- ❏ **Available Resources.** Planners determine the quantity of resources that are available to work on scheduled backlog relief during the approaching schedule week and thereby ensure that commitments are

realistic and do not exceed projected capacity (in any craft or skill set).

- ❑ **Grouping for Optimization.** Planners group jobs that are ready to be scheduled by requesting organization, and equipment to facilitate the effective utilization of asset access and assigned crews by linking multiple jobs on the same equipment, in the same proximity, or requiring the same equipment and special tools.
- ❑ **Ready to go.** The meeting reviews all work orders that are "Ready to be scheduled." This condition means that all requirements are in place and ready except for the assignment of a maintenance team. Once they are in this condition or state, work orders should be filed by required start date. Jobs should not be scheduled if material is not already on-hand, even if the material will presumably be received during the schedule week. There should be ample jobs in the backlog that are truly ready. Except on rare and essential occasions, potential upsets due to late delivery of matereials should not be built into the schedule.
- ❑ The meeting also affords participants the opportunity to request that certain jobs in other status be expedited into the "Ready" state as soon as possible.

Requesting organizations also have responsibility to prepare for the meeting. They should reach accord regarding which of their requested Work Orders are most important for Maintenance to perform during the coming schedule week. It is best if this agreement is reached prior to the coordination meeting so that the meeting can be focused upon reaching accord between, not within, the several requesting organizations.

The give and take contractual process, which the weekly coordination meeting reflects, becomes more binding when managers (operations or maintenance) initially (and occasionally thereafter) chair the meeting. They can pattern it, and "model the way." Once this modeling is accomplished, the Planner/Scheduler can become the chairperson, but operations and maintenance managers ought to attend at least once a month.

As the coordination process is streamlined, the meeting requires less than one hour per week and all participants find that their time and attendance is very worthwhile. Participants realize that the best way to get more work performed properly and in a timely fashion is through the formal planning, coordination, and scheduling process.

Chapter 12

Agenda for the Weekly Coordination Meeting

Item one might be "How did we do against last week's schedule?" and exploration of reasons for non-compliance. What jobs were not completed as scheduled, and why? Did maintenance fail to get to them? Did Operations deny equipment access? Did unscheduled jobs break into the schedule, and were they necessary? Such questions highlight underlying reasons for poor schedule compliance. When the Operations Manager raises the questions the process is more effective, problem resolution is speedier, and future schedule compliance improves rapidly. The Manager should therefore attend the coordination meeting at least once a month.

❑ The preliminary cut of next week's schedule should be shared, showing resources available and demands already established (PM's, Corrective Work, Carry-overs, Project commitments, etc.)

❑ Next, the Production Schedule should be presented to clarify any operational support required from maintenance (set ups, change-overs, change-outs of production expendables, etc.) as well as equipment access windows that can be utilized by Maintenance.

❑ At this point all parties are ready for the give and take as to whose job is next most important in the optimal interest of the overall operation. This continues until all available resources are committed in general terms. Some changes are inevitable when the scheduling process addresses the specifics.

❑ If participants continue to lobby for more support, overtime and contractor support is the next consideration. This requires approval and funding.

❑ Finally, review critical jobs that are delayed for lack of parts, engineering, approval, budget, or other reasons and consider possible expediting.

Scheduling
Maintenance
Work

(THE WEEKLY EXPECTATION)

Once work has been planned, material availability has been assured, and coordination with internal customers has been achieved; work scheduling can be addressed:

Scheduling (when to do the job) is the process by which all resources needed for specific jobs are allocated, coordinated, and synchronized at the proper time and place, with necessary access, so that work can be executed with minimal delay and completed by the agreed upon date, within budget estimates. The schedule establishes when jobs will be done and what resources can best be applied to their performance. Resources include manpower, materials, tools and special equipment. Access refers to when the equipment will be prepared and accessible so that it can be worked on in safe (locked out/tagged out) condition, with necessary precautions taken, permits obtained, and any specialized documentation, drawings, or information in hand. Proper time relates to job start, duration of execution, and completion within the time frame agreed upon with the internal customer during the weekly coordination meeting.

Chapter 13

Scheduling is the locus from which all maintenance activity is executed. When any new maintenance management system is started up, scheduling should be viewed as the "point" function and "marketing arm" of the system because it yields the earliest tangible results (often within weeks of start up).

By contrast, preventive maintenance, equipment history, and maintenance engineering require several months of focused effort before they will yield measurable results. In the meantime, users ask, "What are we receiving from this new system?" Program success demands that this question not be allowed to linger.

All individuals and groups perform better and accomplish more with clearly established, communicated and published expectations. When the maintenance function is managed without a weekly schedule, there are no specific expectations as to what is to be accomplished with the resources for which payroll checks will be drawn. Instead, whatever reactive demands are made is what will be done.

Operational control of maintenance, though much slower to react, is similar to the thermostat principle illustrated in Figure 13.1.

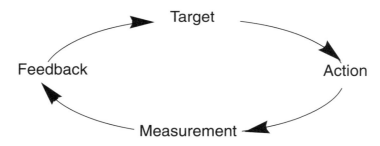

Figure 13.1   The Thermostat Principle

The fundamental requirement target (the schedule) against which to control, followed by action (execution of the schedule) to achieve the target. The results, measured against the original intention (called schedule compliance), provide feedback for correcting deviations. (improving future schedule compliance). Managers must schedule precisely, proceed positively, and persistently pursue weekly targets. Using the results and reports such as those showing schedule compliance, the planner, the supervisor, the maintenance manager, and internal customers can continuously improve the planning, coordination and scheduling function.

Chapter 13

In maintenance, objectives take the form of schedules, or statements of when jobs will be executed. The manager should know specifically what he or she plans to accomplish with each resource, each day of the week.

The overall maintenance challenge is two fold; "to create a maintenance operation which is both responsive to the customer's needs and is intrinsically efficient as well." The schedule is a device for lining up jobs waiting to be performed so that operations are best served while maintenance also makes optimal use of its resources. Four abilities, as listed below, are necessary for maintenance schedules to meet the challenge:

❑ Determine priorities mutual to the involved parties

❑ Focus on the target reflective of those priorities

❑ Concentrate on execution to schedule

❑ Persevere to achieve "Proactive Maintenance Excellence" supportive of World-Class Operation.

One way or another, maintenance resources have a job to be performed. The fundamental question is "are company's resources to be deployed effectively through pro-active scheduling (and the planning that went on before it) or ineffectively consumed through reaction to breakdowns? Urgency alone (without consideration of importance) cannot be allowed to determine how vital resources will be consumed. Excellence is best described as performing the right things properly (by planning), selecting the most important things to be done (by coordination and scheduling) and accomplishing them 100% correctly (by execution) without wasting resources (by planning).

Either you schedule the team yourself, or you allow equipment failure and impatient customers to dictate how and when resources are consumed (and wasted).

The scope of Scheduling includes:

- ☐ Bringing together in precise timing the six elements of a successful maintenance job: labor; tools; materials, parts and supplies; information, engineering data and reference drawings; custody of the unit being serviced; and the authorizations, permits, and statutory permissions.
- ☐ Matching next week's demand for service with resources available after accounting for all categories of leave, training, standing meetings, and indirect commitments, plus consideration of individual skills.
- ☐ Preparation of a "Weekly Schedule" that represents the agreed upon expectation regarding planned work orders to be accomplished with available resources. The schedule also assures that all preventive and predictive routines will be accomplished within established time limits.
- ☐ Consideration of alternative assignment strategies where the schedule assigns specific jobs to specific people (allowing second-string players into the game to gain experience … as feasible).
- ☐ Ensuring that responsible supervisors receive and understand the planned job packages for scheduled jobs.

Schedules encompass periods of full production, as well as shutdown and partial shutdown periods (down days, major outages, as well as the annual turn-around). Very large jobs or shutdowns require independent, critical path schedules because of the number of activities to be executed.

BACKLOG MANAGEMENT IS A PREREQUISITE
FOR EFFECTIVE SCHEDULING

Effective scheduling requires adherence to proven principles of backlog management and established procedures:

- ☐ Lead-time – work to be done must be identified as far in advance as possible so that the work backlog is known and jobs can be planned effectively and completely prior to scheduling.

❏ Accurate evaluations should be made of the priority of each job, given the perspective of the overall operation. Each job in backlog must be force ranked so that the most important jobs are always scheduled and where possible, executed first.

❏ Backlogs must be kept within a reasonable range. Backlogs below minimum do not provide a sufficient volume of work to ensure smooth scheduling and effective utilization of all resources. Backlogs above maximum turn so slowly that it is impossible to meet customer needs on a timely basis. Consequently, they loose faith in the proactive approach and slip back into the reactive quagmire.

❏ Special or heavy demands resulting in excessive backlog cannot be scheduled unless additional resources are authorized or expected completion dates are relaxed.

Adherence to these prerequisites ensures that all maintenance needs are properly attended, customers have their work performed on a timely basis, equipment experiences minimum delay, work is performed safely, and overall maintenance cost is kept to an optimum. The summation of these benefits combines to reduce the overall cost of a quality product.

SCHEDULING TECHNIQUES

The scheduling process establishes a plan for the effective utilization of all maintenance resources in the interest of reliable operating capacity. The two common techniques are outlined below:

❏ GANTT Bar Chart — As shown in Figure 13.2, this technique shows the time relationship of job tasks in terms of their chronology and simultaneousness. Such charts are useful, but do not convey task relationships, because it is not clear which activities must finish before others can begin. This is the technique to use for weekly maintenance scheduling. As such it is the focus of this chapter.

❏ Network Arrow Diagrams — This technique, depicted in figure 13.3, takes two basic forms; Critical Path Method (CPM) or Project Evaluation and Review Technique (PERT). By identifying the "critical path", both these forms depict the shortest elapsed time feasible for completion of major projects. The importance of scheduling in this manner increases as project

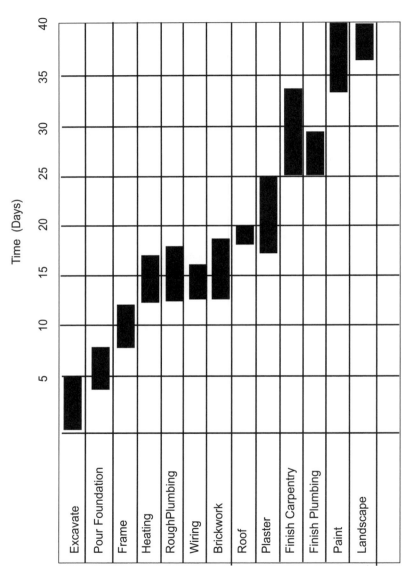

Figure 13.2   GANTT Bar Chart

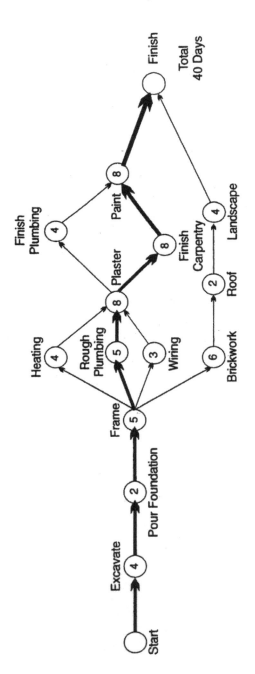

Building a House on the Critical Path

Figure 13.3   Network Arrow Diagram

complexity increases andon-time completion within budget constraints become critical.

This unique form of planning, scheduling and control is covered in Project Management (Chapter 18).

INSTRUCTIONS FOR PREPARING SCHEDULES

Experience shows that increased productivity is achieved when maintenance personnel know tomorrow's assignments before they leave work today. Knowing about work in advance enables a craftsman to make mental preparation. With good scheduling, the maintenance supervisor is provided with a list of all work to be performed by members of his team, both individually and as a unit. Every evening, the supervisor can tell each team member his or her assignments for the next day.

Scheduling principles are simple, but their effective application is not quite so easy because three separate procedures must be performed concurrently to produce a viable schedule. These procedures are:

❑ Job loading, to set forth jobs to be completed during the schedule period. Scheduling of these jobs stems from coordination between maintenance and operations to assure that the near term operating needs and the long-term assurance of asset and capacity reliability are both served.

❑ Job scheduling, to sequence the loaded jobs through the schedule week based on meaningful estimates of the required duration and agreed upon equipment access.

❑ Manpower commitment, to ensure optimal utilization of resources.

An effective schedule, therefore, consists of three sections corresponding to the three procedures, as shown progressively in Figures 13.4 through 13.6. When completed the schedule provides the expectation of accomplishment for a full week... with resources allocated by job, by day, and by individual maintenance technician or crew (in the case of multi-person jobs).

134

The following scheduling guidelines and techniques for planning job loading are recommended.

JOB LOADING

Jobs should be listed in sequence of agreed upon importance during the weekly coordination meeting. This sequence may not be chronological within the week. This most important job might be scheduled at the end of the week to accomodate production needs and their ability to release the asset.

Line Item	WO #	Job Description	Req. Skill	Crew Size	Clock Hours	Man Hours	Assigned Crew
1	10010	Daily PM Line #2	M3	1	1	1	F
2							
3							
4							
5	10340	Replace Limit Switches	E4/E1	2	4	8	A & B
6							
7							
8		Repl. Limit Switches - Ln #1	E4	1	2	2	C
9							
10							
11	10362	Connect Power to Line #3	E1/E4	2	4	8	C&A
12	10362	Check Run Line #3	M1/M3	2	4	8	G & F
13							
14							
15							
16							
17							
18							
19							
20	10020	Clean & Insp Load Centers	E4/E1	2	4	8	A& B
			Total Man-Hours Scheduled:				
Add 10% Provision						"A"	

Table title: Job Loading

Figure 13.4   Job Loading

JOB SCHEDULE

The careful selection of your most logical schedule week will contribute significantly to schedule compliance. Notice the unusual sequence of the schedule week (Friday to Thursday) in Figure 13.5. Commonly, when Maintenance and Operations meet, they first discuss work that needs to be completed over the weekend. These discussions progress to preparatory work that can be performed on Friday thereby avoiding use of narrow access time for fabrication and other preparatory work. While a schedule week beginning on Friday and ending on Thursday makes sense in such situations, it may not be best for your environment. You need to determine what would be best for your operational environment. In making the determination, consider when the coordination meeting would be held and whether Operations has their schedule adequately in hand by that point of the week Scheduling and schedule compliance will be off to a much stronger start if the schedule week makes practical sense to all involved parties.

When laying out the schedule, the important considerations are the duration of the job, when it is to be performed (start and finish), and what crew member(s) should be assigned to it. Start and finish dates and times are particularly important when shutdown of production processes and coordination of multiple resources are involved. A typical job schedule is added to Job Loading in Figure 13.5.

In that figure, each square represents one hour, and the time-line is blocked to show scheduled start, duration, and finish. The form illustrated provides for shifts of up to twelve hours including overtime. As each job is tentatively aligned to a specific team member or members, the letter code of each is indicated in the "assigned to" column, thereby cross references jobs to human resources.

Using a recognizable key, insert notes at the ends of bars and at the beginnings of succeeding bars where shutdowns and support crafts are required (electricians must complete line item 11 before mechanics can begin line item 12).This requirement reminds supervisors that deviation from schedule on these jobs dictates notification of the impacted planner, supervisor, and requestor. Planners must further coordinate their plans directly to ensure that all schedules are synchronized.

Only a few jobs are illustrated here in the interests of simplicity and clarity, but, of course, the schedule should be fully developed.

LABOR DEPLOYMENT SCHEME

As jobs are aligned to individuals on the Job Schedule, those individuals must be committed to their assigned jobs within the Labor Deployment Scheme (Figure 13.6). Thereby cross-referencing resource to

Job Loading

Line Item	WO #	Job Description	Req-Skill	Crew Size	Clock Hours	Man Hours	Assigned Crew	Friday	OT	Saturday	Thursday	OT	Completed Man-Hours
1	10010	Daily PM Line #2	M3	1	1	1	F						
2													
3													
4													
5	10340	Replace Limit Switches	E4/E1	2	4	8	A & B						
6													
7													
8		Repl. Limit Switches - Ln #1	E4	1	2	2	C						
9													
10													
11	10362	Connect Power to Line #3	E1/E4	2	4	8	C&A			M			
12	10362	Check Run Line #3	M1/M3	2	4	8	G & F			E			
13													
14													
15													
16													
17													
18													
19													
20	10020	Clean & Insp Load Centers	E4/E1	2	4	8	A& B						

Total Man-Hours Scheduled: "A"

Add 10% Provision

Man-Hours Completed: "B"

Jobs On Contingency Schedule

25													
26													
27													

Figure 13.5 Job Schedule

assigned job.

A master schedule template similar to that shown in Figure 13.6 should be established for each crew or team responsible to a specific supervisor or team leader (approximately ten people). Personnel comprising the team should be pre-posted to the template.

Should any team resource have a fixed commitment to indirect activities such as training, the pre-committed capacity should be indicated on the master template (see employees K and L on Wednesday afternoon). Scheduling of committed capacity to a job at a time when the resource is not available for direct assignment is thus avoided. Such things are easily forgotten at inopportune times.

The time periods comparable to those to which a job is scheduled are shaded alongside the appropriate individual's names. . This procedure minimizes the risk of scheduling an individual to two jobs at the same time, another situation that occurs from time to time.

Each assignment should be cross-referenced to the appropriate job. In the interest of conserving schedule space, reference should be to the line item number (one or two digits), rather than the work order number (four or five digits). When the schedule is completed, all resources available for the week should be committed.

❑ When all three sections are brought together on a single form, (figure 13.7) a complete picture is provided for the coming week that assures: Nobody is scheduled to more than one job at any given time

❑ Available resources are fully utilized without voids or overloading

❑ Internal customers receive the promised support

When schedules are not detailed by individual, but taken only to the level of a full load of identified work for a given team or skill, too much work tends to be loaded for some and too little for others. Individual team members are not clones of each other, even if they are multi-skilled. When uneven loading is allowed to happen, the resultant schedule is not feasible from the beginning.

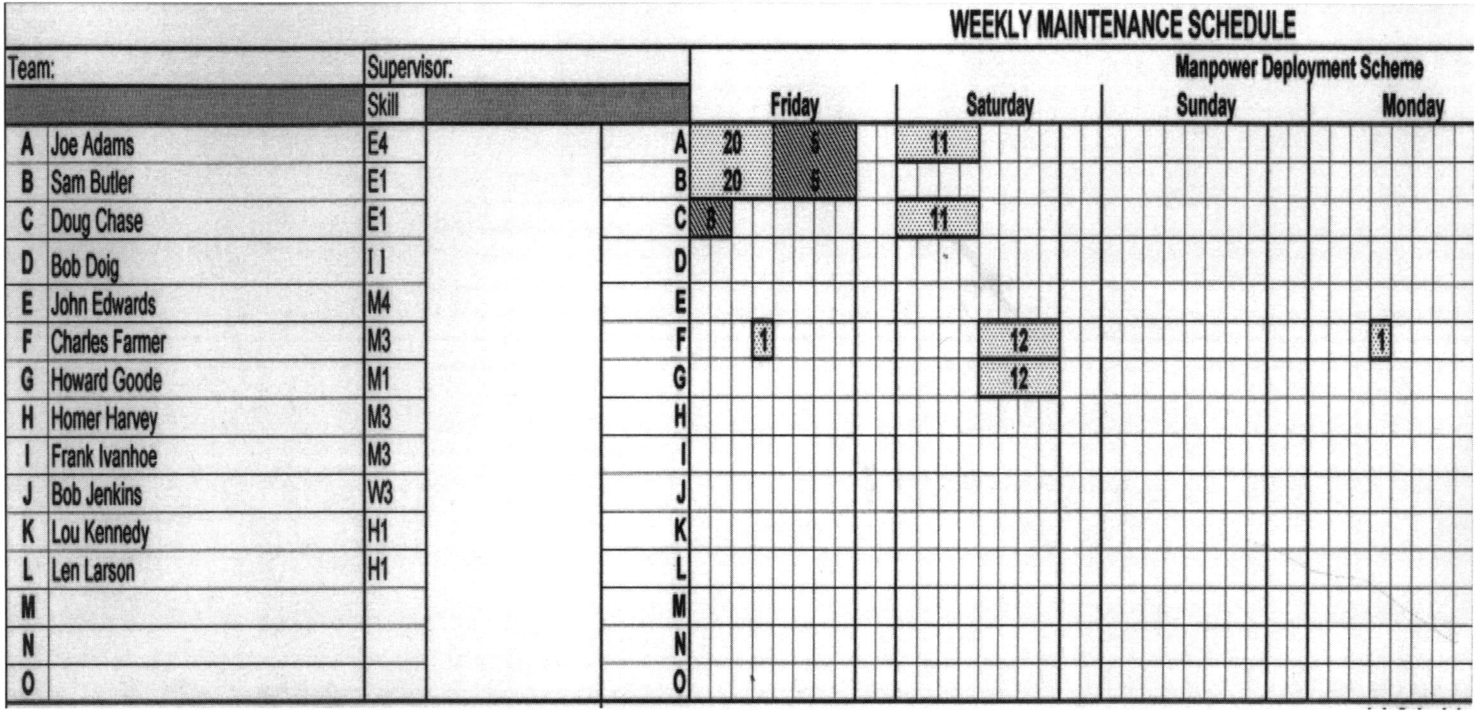

Figure 13.6 Labor Deployment

WEEKLY MAINTENANCE SCHEDULE

Complete Scheduling Format

Manpower Deployment Scheme — Week Beginning:

Team:	Supervisor:		Friday	Saturday	Sunday	Monday	Tuesday	Wednesday	Thursday	Completed Man-Hours
		Skill	20	11						OT
A	Joe Adams	E4	20	11						
B	Sam Butler	E1								
C	Doug Chase	E1								
D	Bob Doig	I1								
E	John Edwards	M4								
F	Charles Farmer	M3	1			1	1	1	1	
G	Howard Goode	M1								
H	Homer Harvey	M3		12						
I	Frank Ivanhoe	M1		12						
J	Bob Jenkins	W3								
K	Lou Kennedy	W3						T T T T		
L	Len Larson	H1						T T T T		
M		H1								
N										
O										

Job Loading

Line Item	WO #	Job Description	Req. Skill	Crew Size	Clock Hours	Man Hours	Assigned Crew
1	10010	Daily PM Line #2	M3	1	1	1	F
2							
3							
4							
5	10340	Replace Limit Switches	E4/E1	2	4	8	A & B
6							
7							
8		Repl. Limit Switches - Ln #1	E4	1	2	2	C
9							
10							
11	10362	Connect Power to Line #3	E1/E4	2	4	8	C&A
12	10362	Check Run Line #3	M1/M3	2	4	8	G & F
13							
14							
15							
16							
17							
18							
19							
20	10020	Clean & Insp Load Centers	E4/E1	2	4	8	A & B

Total Man-Hours Scheduled: "A"

Add 10% Provision

21							
22							
23							
24							

Jobs On Contingency Schedule

25	
26	
27	

Job Schedule

Assigned Crew	Friday	Saturday	Sunday	Monday	Tuesday	Wednesday	Thursday	Completed Man-Hours
		OT	OT	OT	OT	OT	OT	OT
F	□			□	□	□	□	
A & B								
C		M / E						
C&A								
G & F								
A & B								

Scheduled Man-Hours Completed "B"

Pages can be added to the Job Loading and Job Scheduling sections (lower portion of Figure 13.7, but the Labor Deployment Scheme must be worked and be visual in the one location (upper portion of the first page).

COMPLETING THE SCHEDULING PROCESS

The following scheduling guidelines and techniques are offered

❑ Prepare a schedule form for each supervised unit by entering the week beginning date, the name of the responsible foreman and the organizational unit involved.

❑ The *Planner* should determine (by discussion with the foreman, reference to vacation charts, and other means), known absences for the coming schedule week and subtract them from the resources expected to be working during the schedule week.

❑ Review all jobs in the backlog starting with incomplete jobs from current or previous schedule(s) .

❑ Review **Planned Job Packages, to** make certain they are sufficiently complete for scheduling and assignment. This work includes final confirmation of material, parts, and special tool availability required for execution, safety instructions, and permit requirements.

❑ Plan strategy on a weekly basis. Rigidly enforce the rule that weekly schedules must be prepared for each supervisor by Thursday Noon of the preceding week (if the schedule starts Friday). The schedules are to show how team resources are to be utilized throughout the schedule week.

❑ Work scheduled must be balanced against available man-hours, and sufficient jobs must be posted to the schedule to consume all available labor-hours. Schedule what can be done, not necessarily what needs to be done.

❏   Each available mechanic should be scheduled for a full day of productive work for each day of the schedule week. The duration should be indicated in the job section of the schedule. In the man power deployment section, indicate labor-hours, to avoid confusion.

❏   The majority of the crews should be scheduled for important work, which needs to be started and completed without interruption. Make a conservative provision for urgent schedule breaks. Assign jobs that can be interrupted or delayed, to "a few good people" who are flexible. Flexible means that they can stop and resume jobs, be re-instructed and reassigned to "emergencies" with minimal loss of efficiency and without a drop in morale. Approximately 10 to 15% of scheduled labor-hours should be on low priority jobs that can be sacrificed when necessitated by urgent demands. Personnel assigned to such jobs are the ones to be pulled in response to schedule breaks (urgencies).

❏   Do not schedule any job until all needs (parts, materials, tools, special equipment, asset access, the item to be worked, any special support) are available in the quantity required and at the time necessary.

❏   List jobs in descending order of importance until all available man-hours are committed (PMs listed first), based upon agreements reached during the weekly coordination meeting.

❏   Determine most logical time of day to schedule PPM's. Often, the early part of the day is the period of heaviest breakdowns so is not a good time to schedule PM's. On the other hand, it is not advisable to leave them until the end of the day because then they may not get done at all. Late morning or early afternoon are advisable times to schedule PM's (as shown on figure 13.7-Line Item 1)

❏   Add jobs equal to 10 or 15% of scheduled labor-hours (Line Items 21–24) as provisional jobs to be substituted when scheduled jobs are unavoidably delayed or completed in less than the estimated time.

❏   Establish a contingency section of the schedule (Line Items 25-27) for jobs of high desirability, but that require equipment access not expected within the schedule week. Should availability occur, it is more important that these jobs be performed than some jobs on the primary schedule, but only if the provisional jobs have been properly planned. The provisional strategy is proactive and should be classified as schedule compliant.

❏   Avoid duplicate shutdowns by scheduling all work requiring common equipment access as appropriate.

❏   Save minor indoor jobs for severe temperatures and inclement weather.

❏   Eliminate unnecessary trips. Look for opportunities for assignments to take advantage of jobs in the same location, jobs using the same tools or materials, jobs needing the same skills, and other ways to improve efficiency.

❏   Schedule multi-person jobs as the first job in the morning whenever possible so that everyone is available to start the job at the same time.

❏   When scheduling multi-person jobs later in the day, consider previous assignments. Don't assign one person to a one-hour job and the helper to a two-hour job because both will not be available to start the two-person job concurrently.

❏   Think about crew balancing delays on multi-man jobs. All four members of a crew are seldom required for the entire duration of a job. Often another small job in the same area can be worked con currently.

❏   Allocate people to specific jobs with supervisor's approval. Pick the people for the jobs based on knowledge and aptitude, required skill or equipment and on the basis of individual training needs. Experience shows who is skillful in certain job types and who needs more exposure to them. Balance equipment specialization with broad facility knowledge. Utilize individual skills to the greatest extent possible. Craftsmen should be provided with a challenging environment and the opportunity to grow.

❑ Schedules for the forthcoming week for each supervisor's team must be finalized and posted prior to the end of the previous week. All preventive and predictive maintenance inspections must be incorporated at their predetermined frequencies and the timely completion of all identified corrective maintenance must be scheduled.

❑ Associated "Planned Job Packages" must be delivered to and reviewed with responsible supervisors to assure that nothing falls through the cracks due to misinterpretation of intent or meaning. These consultations form the point at which responsibility transitions from planner to supervisor. Nothing can be allowed to be lost within the transition. In turn, the same level of transition must take place between supervisor and technician at the time of assignment.

❑ Operations are to be provided with copies of schedules to confirm and document that all agreed upon commitments are acceptable and understood by both operations and maintenance departments.

❑ It is vital that schedules be studied and approved by everyone concerned. Approval means that a contract has been reached between operations and maintenance to comply with "their joint schedule" for the deployment of maintenance resources in support of operating plans.

❑ At this point, the Weekly Master Schedule becomes a document of which all parties, through mutual contribution, accept ownership.

❑ When urgent work is done at the expense of scheduled jobs, a schedule overload results. A scheduled job will be displaced and carried over to the next schedule period, unless the problem is addressed by a temporary increase in capacity (overtime or contract labor). The displaced job is one of those scheduled for the organization that initiates the schedule break. Therefore, requests for schedule breaks require the sanction of the Maintenance and Production Manager.

❑ Finalize tactics on a daily basis when the schedule is being executed. The weekly schedule must be updated each evening during the week it is in force for the balance of that week. While the transition from reactive to proactive maintenance is taking place, updating will be burdensome and will have to be performed by the planner. However, as schedule compliance matures, the required updating becomes minimal and can be performed by the supervisor.

❑ Operations must advise planners at the earliest possible moment if they are unable to release equipment as scheduled. Similarly, the maintenance department must advise production management if the reverse situation is likely to occur. Planners must ensure the coordination.

❑ Planners must keep abreast of schedule status, and detect when a job runs into trouble before it misses a milestone.

❑ Maintenance must notify and consult with customers about any pending interruptions or disruptions.

❑ When a job is complete, maintenance must collect the planned job package with appropriate feedback, record the results for schedule compliance, and confirm that the job is closed out. Feedback includes what actually happened, what failed, and ideas for improvement.

❑ Maintenance must verify that the job was done according to the plan. When a job deviates it is vital to learn why. Verify that the job used the materials listed in the Bill of Material. Verify that all specialized tools and equipment were accounted for in the plan. Verify that drawings were correct and that no additional permits or permissions were needed. Finally, on larger or disruptive jobs verify that all people who should have known about the job were notified and all processes were shut down appropriately.

❑ Finally, update the planning package in all the areas mentioned.

Job Execution

The planned jobs are prepared. The schedule is complete. Monday morning (or whenever your schedule starts) rolls around. The ball shifts to the supervisor's court. The supervisor becomes responsible for tactical execution of the schedule.

THE SUPERVISOR IS RESPONSIBLE FOR JOB EXECUTION

The supervisor has to be in several places at once. He or she has to keep ahead of jobs that are about to start and periodically check in on jobs that are already in process. In this manner, the progress of each job relative to the schedule is continually reviewed to determine if the situation has changed in any material way.

This requires timely information to determine when projects and other jobs are progressing unfavorably. An accurate schedule helps supervisors to judge when exceptions fall outside of reasonable boundaries and intervention is in order. Too many supervisors try to stay abreast mentally or in a pocket pad. A comprehensive schedule is much easier and extends the ability of all supervisors.

Effective supervisors also assure that team members reference their planned job packages to minimize exceptions before they occur. The packages have been prepared for good reason. Use them!

Staying abreast of job status relative to schedule enables the supervisor to take corrective action before problems become serious. When a time-line is included within the planned job package, the supervisor can make a well-informed judgment. If the schedule shows the job should be half done by lunchtime, the supervisor quickly looks for a certain amount of work to have been completed. If the amount of work completed is within reasonable boundaries, no intervention is needed. However, if the supervisor finds that an important job is falling behind schedule, tactical decisions (larger crew, overtime, contractor support) can be made to correct the situation before a major shortfall develops. Intervention while the job is still in process can make a real difference in downtime and equipment availability.

If thirty minutes after job start the crew is still looking for a ladder to begin the job, the supervisor should jump into action. When jobs go awry, materials, parts, and tools are the most common culprits. Obviously, there were shortfalls in the planned job package, provide feedback to the planner. Feedback after job completion enables the planner to improve the planned job package, thereby precluding such delays in the future.

The above scenario reflects true management. Explanation of variances after the fact (horse out of the barn) does not.

THREE IMPORTANT FUNCTIONS

❑ During the schedule week, the supervisor checks preparedness for each day's work. Despite the planner's preparation, checking all the resources listed in the planning package and situated at the job site assures effective execution.

❑ Supervisors make the actual job assignments to specific team members. At times, he must make tactical decisions about the assignments, but in most cases he follows the labor deployment scheme provided in the upper portion of the Weekly Schedule (Figure 13.7). The planner has put forethought into the balance between aligning the job to the 'best' possible tradesperson or to the tradesperson who needs the experience the most.

Chapter 14

DAILY SCHEDULE ADJUSTMENT

The schedule requires adjustment after each day, to provide a recasting for the balance of the week. Ultimately this recasting is a supervisory responsibility but, until the transition from reaction to proaction is well underway, the planner/scheduler will probably need to do it. Planners should get beyond this stage as quickly as possible. Until they do, they will spend too much time on scheduling to the detriment of planning.

During daily schedule adjustment, the planner (initially) Supervisor (ultimately):

➢ Follows schedule progress and coordinates schedule adjustments as dictated by "real" operating needs and changing resource constraints. Urgent interruptions will occur regardless of how well schedules are constructed and coordinated, and will change the framework of work for the week. Supervisors must therefore review, adjust, and update their schedules as necessary each evening for the remainder of the schedule week to coordinate new high priority work orders with those already on the weekly schedule.

➢ Supervisors should always strive to optimize schedule compliance, despite essential "schedule breaks." A description and note of the duration of the "schedule breaks" (jobs that interrupt the schedule) is to be included at the bottom of the schedule. When schedule compliance does falter, every effort must be made to recover from interruptions and protect as much of the original schedule as possible.

➢ Adjustments to schedules should always be made in such a way that the most important jobs are completed by end of the week. Because scheduled jobs are listed in order of importance, the jobs to be protected from interruptions are always in the forefront. Extended or delayed jobs must be carried over to the next day … particularly if they are already started. Schedule adherence or protection is usually a cost effective use of overtime. Selected scheduled jobs should sometimes be sacrificed to take advantage of downtime windows of opportunity to perform planned jobs listed on the contingency schedule.

➢ Team members must always be kept fully aware of alterations in assignments and their timing, and Production supervisors also must be kept abreast of revisions. Some organizations enforce a rule that supervisors cannot make schedule changes without defined authorization.

PLANNER SUPPORT OF JOB EXECUTION

❑ *Support of Job Execution*— ensures that the responsible supervisor receives and understands the planned job package for each scheduled job.

❑ Follow-up coordination ensures that all agreed-upon supportive actions of others are performed as planned.

❑ *Schedule Follow-up*— determines the level of schedule compliance and reasons for non-completion. This "constructive"responsibility works toward future improvement. At the end of the schedule week, Superintendents should review schedule compliance with all supervisors for whom they are responsible.

THE MORNING MEETING

A well-thought-out schedule provides a framework for achievement of weekly targets, but problems do occur and changes need to be made on a daily basis. All parties must be involved in discussions related to course corrections. The morning meeting is such a forum and is often effective, especially when the maintenance function is still operating in a reactive environment.

Once the morning startup is accomplished, and inputs that must be factored into the schedule have been gathered from the previous evening and night shifts, the maintenance superintendent holds a short meeting to refocus the organization upon unified targets.

The meeting focuses on immediate problems while protecting as much of the weekly schedule as possible. Tactics are therefore discussed and resources realigned as necessary.

The meeting is short, perhaps 15-30 minutes. Therefore, it is often conducted standing up. The focus should be on getting the desired result and not on who made the mistake. An upbeat attitude, oriented toward problem solving, is the tenor of the meeting.

Everybody leaves the meeting knowing the joint priorities for the day. If schedule change is needed before the next meeting, the superintendent is informed. Only he or she can change the schedule between morning meetings.

Once the cultural transition from a reactive to a proactive environment has been achieved, emphasis is switched to the weekly coordination meeting between maintenance and operations. The daily meeting then has less prominence and superintendent approval becomes a requirement prior to any change of the weekly schedule.

15

Job Close Out, Follow Up, and Schedule Compliance

Job close out begins with feedback from craftsmen and supervisors. The job is not complete without comprehensive feedback regarding the work performed. In the case of planned jobs it need be only to exceptions, additions and deletions from plan. If planning was effective, such need should be minimal.

The most basic feedback is labor hours charged, which should be reported via the Labor Distribution System. The second basic is any materials and parts, which the team had to requisition themselves. This should be captured via the Stores Requisition system. Feedback on the Work Order itself would be expanded or corrected description and job steps; additional special tools required, and interruptions encountered, etc.

Upon job closure, planners have an opportunity to acquire an excellent snapshot of what it takes to perform a given job (not necessarily actual, if performance was poor) and to add that knowledge to the building of a valuable library of Planned Job Packages. This opportunity presents itself every time a job is completed. Historians say, unless we study history we'll be doomed to repeat it. Maintenance jobs are similar. Thousands of dollars are spent doing a particular job, but it seems to be difficult to spend even $50 to close it out in such a way that we can take advantage of the detailed knowledge gained from doing the job in the first place. During job close out and follow up, the Planner (with administrative support) has several responsibilities. One of the most important is feeding good data to the CMMS. It is essential for future analysis that garbage is kept out and details are provided. Garbage takes many forms. A work order that states, "Fixed pump," 2 hours, is an example.

Data concerning urgent jobs must also be fed into the CMMS in a complete and accurate manner. Breakdown information is the most important part of Equipment History, as it is this work that requires Root Cause Analysis.

A detailed and complete Work Order for a breakdown might form the basis of a job pre-plan (terminology for a planned job package covering emergency breakdowns). Ultimately, even repairs of emergency failures can be planned. We know what failures can occur. We just don't know when. With the establishment of Job Pre-Plans, we are ready whenever they occur.

In performance of his close out responsibilities, the assisted planner must:

❑ Collect completed work orders and enter them or direct them to be entered into the CMMS. He or she must ensure that all fields are filled in, comments are readable (contact the writer if not) and that there is enough detail to support future analysis.

❑ Close work orders, as soon as completed, to equipment history and charge them to the proper accounts, thereby avoiding inappropriate charges after jobs are complete. The distribution should of course be electronic. Review use of labor and materials relative to estimated usage, to determine what refinements of the planned job package are necessary.

The planner also verifies that all essential feedback entries have been made and schedule breaks are recorded at the bottom of the schedule.

Many jobs have leftover materials and parts. The planner is the point person in ensuring that materials and tools are properly returned. Some care should be taken over this point because maintenance craftspeople, particularly in reactive environments, tend to be packrats. In some shops you never see leftovers because they are squirreled away somewhere (maybe never to be found again). The planner has to check proper disposition of leftovers under the following heads:

- ❑ Equipment (should be handled by the Planner). If the equipment is rented the planner insures that it is returned, to avoid excessive charges.

- ❑ Purchased materials and parts (by Planner). Where possible these materials or parts should be returned to the vendor for credit. If they cannot be returned they should be stocked if there is some continuing value for them, or scrapped if not.

- ❑ Stock items (Supervisor or craftsperson is responsible). The stock keeping system should have an easy way to return stock and get a credit for it against the work order.

- ❑ Special tools (Supervisor) are returned to the tool crib. For rented tools, see Equipment.

- ❑ Free Bin Stock (Craftsperson)

SCHEDULE COMPLIANCE

The maintenance supervisor keeps the planner abreast of schedule compliance throughout the course of the schedule week. The planner notes status of individual scheduled jobs, and produces a summary review at the end of each week.

The planner calculates Schedule Compliance ratios each week and prepares associated reports. The calculations determine the level of schedule compliance and reasons for non-compliance, not to place blame but to identify constructive actions to improve future performance.

The schedule is not rigid, but represents the most desirable target for the supervisor to pursue. At the same time, supervisors must have flexibility to make necessary tactical decisions as true urgencies arise.

Non-compliance is especially serious when Maintenance fails to bring capacity back on-line as promised. This failure stands in the way of cooperation and building the participative partnership. It also feeds distrust and future reluctance to release assets for scheduled maintenance.

REASONS FOR NON-COMPLIANCE

When scheduling is first launched, many previously hidden problems surface. Schedule compliance highlights areas where mechanics cannot do their job due to problems outside their control. These reasons should be recorded, reported, and studied for trends. The following coding structure is offered as an example.

Figure 15.1 Reason for Schedule Non-Compliance	Suggested Code
Operations may fail to release equipment as previously agreed to.	FR
Excessive emergencies. Technicians are pulled off scheduled jobs to work on less important jobs without consultation or communication with maintenance leadership. (Commonly found where operating personnel are allowed to redirect maintenance personnel rather than going through the work management process). Poor assignment of technicians. Completions are bound to fall short of expectations if schedules are not feasible from the outset. Schedules often are not feasible when they are not detailed to the level of individual assignments.	PA EE
Insufficient technician capacity. Lack of cross training contributes to shortage of essential resources.	IC
Stock-outs are frequent. Inaccurate inventory control regularly contributes to compliance shortfalls. Quantities shown to be in inventory are not correct, and excessive time is spent at the storeroom because required materials and parts have not been pre-requisitioned and delivered to the job site.	SO
Planning packages do not reflect reality. Parts are incorrect, job steps are incomplete or wrong, and lockouts, specifications and regulations are not documented.	PP
Failure to meet estimated job durations means that some scheduled jobs are not completed by end of the scheduled week. The immediate conclusion is that the schedule must be wrong. Planners must not jump to the conclusion that the estimate was wrong. It may have been, but the crew may have performed poorly, or another identified problem may have occurred.	DU
Excessive absenteeism and simultaneous peak loads.	EA

Chapter 15

CALCULATION OF SCHEDULE COMPLIANCE

The scheduling format presented in Chapter 13 provides two columns for collection of the necessary information to calculate "Schedule Compliance." The formula is:

Schedule Compliance= B/A (on scheduling format)=
Scheduled Labor-Hours Completed / Scheduled Labor-Hours Scheduled

The right column of the format provides the numerator (B). The fourth column from the left provides the denominator (A). Use the cells labeled "B" and "A." "A" is used as the denominator because it represents all the work scheduled and expected to be accomplished, and it is not inflated by either the 10% provisional or by the contingency schedule. "B" is used as the numerator and includes work substituted and completed from the provisional or contingency schedules.

If 1000 hours were scheduled (that would be 'A') for the week and 800 were completed (B), the Schedule Compliance would be 80%.

SUPPLEMENTARY METRICS

Two additional metrics relate to scheduling (Schedule Performance and Schedule Effectiveness).

Schedule Performance = Scheduled Labor-Hours Completed / Total Labor-Hours Available

In the above example, if 800 labor-hours of scheduled work was completed but there were 1600 labor-hours available (after vacations and absenteeism), then:

Schedule Performance = 800 scheduled hours completed/1600 hours available = 50%

Schedule Effectiveness = Schedule Performance x % of Labor-Hours Completed that were direct versus indirect work

Effectiveness is the most complete metric. . In the above example, if only 600 of the 800 hours of scheduled work completed were direct versus indirect work, then: **Schedule Effectiveness = 50% x 0.75 = 32.5%**

Chapter 15

The three metrics combined provide a comprehensive picture of scheduling success:

❑ Schedule Compliance measures the percent of scheduled labor-hours completed during the schedule week.

❑ Schedule Performance measures the percent of labor-hours available to be scheduled that were indeed scheduled and completed during the schedule week.

❑ Schedule Effectiveness measures the percent of total labor-hours worked that were scheduled to direct work (vs. indirect work) and completed during the schedule week.

Under no circumstance are Schedule Breaks (urgent jobs that broke into the schedule) credited to any of the three schedule compliance metrics. However, the breaks are credited when Crew Efficiency is calculated. Efficiency is almost always higher when a job has been properly planned and scheduled.

The three metrics are needed because:

❑ Compliance can be high simply because only a small portion of departmental capacity was scheduled. Performance will highlight this shortfall.

❑ Compliance and Performance can both be high, but can be inflated by a lot of indirect work that was added to the schedule, i.e., training, infrastructure assignments, loaned to engineering, and so on. If this type of assignment does not appear on the schedule, Effectiveness is not needed. But if indirect work is added to the schedule, Effectiveness becomes essential. Indirect assignments are practically automatic compliances (100%) and drive the composite compliance and performance up accordingly.

16

Planner and Scheduler Metrics

Measurement of maintenance effectiveness requires the use of a family of metrics with related benchmarks (a few dozen in total). Planning and scheduling impacts a majority of these metrics and planners are responsible for administering many of them, including information assembly, calculation, trending, and posting. However, only a handful can act as measures of a Planner's own efforts.

For instance, Crew Efficiency and Schedule Compliance are not measures of planner effectiveness but are measures of maintenance supervisors, their crews, and production management. These measures tell (in part) whether timely access to the equipment on which maintenance work was to be performed was denied, or whether the equipment was ready, in the agreed upon state, for the maintenance work to be done.

The function establishing the expectations (estimates and schedules) will not retain independent objectivity if it is measured by the same metrics as the supported supervisors and crews. Situations often occur where performance reviews of planners are based upon how closely actual man-hours equal estimated man-hours and/or how close Schedule Compliance comes to 100% (see Chapters 12 and 13 for the calculations). The planner's job is not to predict results but to establish expectation (superficial goals lead to superficial results). This reflects the difference between a Historical Average and a "Standard", as discussed in Chapter 11.

Planners must always aim high. Expectations influence outcome and create transformations in people, Pursue things that will make a difference rather than seeking the safe path of mediocrity.

DIRECT MEASURES OF PLANNING EFFECTIVENESS

Most of the measures listed below should be looked at monthly and plotted on a chart to allow trends to be detected. The most meaningful and appropriate measures of planner effectiveness are:

- ❏ Percent of Work Orders covered with Planned Job Packages
- ❏ Percent of Work Orders covered with an Estimate of Required Labor-hours
- ❏ Reliability (accuracy, currency and completeness) of Backlog by Status
- ❏ Mean Time from Work Order Request to Ready To Be Scheduled Status (other people and functions also influence this metric)
- ❏ Mean Time between Job Completion and Job Close Out
- ❏ Steady and meaningful expansion of Planner Reference Libraries
- ❏ Customer Satisfaction with Planner Communication, Coordination, and Feedback
- ❏ Supervisory Satisfaction with Planner Support on Plannable Work Orders (periodic survey)
- ❏ Crew Satisfaction with thoroughness of Planner Preparation for the smooth execution of planned jobs. Thoroughness can be measured by application of "the Job Plan Survey" included among the appendices (periodic survey).
- ❏ Timely posting and distribution of Weekly Schedules together with distribution of associated Planned Job Packages
- ❏ Timely and accurate preparation, distribution, and/or posting of those control reports and trend charts for which the planner is responsible

Chapter 16

INDIRECT MEASURES OF PLANNING EFFECTIVENESS

All other measures of planner contributions are qualitative measures of maintenance functional performance, the credit or criticism for which belongs to the entire maintenance department and indeed to the entire local organization. Among these measures are:

❑ Improved plant conditions
❑ Improved use of labor as measured by a reduction of emergencies to a goal less than 10% of maintenance manpower, and the management of maintenance overtime within established control limits. The authors' recommendation is between 7% and 15% of straight time on a year-to-date average.
❑ Satisfaction of required job completion dates as specifically requested or implied by agreed upon priority
❑ Improvement in Mean Time Between Failures (MBTF). This is also a primary measure of Maintenance Engineering effectiveness.
❑ Management of backlogs within specified control limits (Ready between 2 and 4 weeks. Total between 4 and 6 weeks). Management commitment is a major factor to the preservation of these control limits.

THE FOLLOW-UP CRITIQUE

Planning quality is best measured by post completion feedback and critiquing.

This process should take place on an ongoing basis as an agenda item within the Weekly Coordination Meeting. The first subject on the agenda must be critiquing of the recently completed schedule before any discussion of the upcoming schedule. This weekly meeting is the most timely and dynamic opportunity to evaluate the quality of planning, measuring, procurement, coordination, and scheduling. Questions such as the following should be addressed:

❑ Was the schedule successfully completed?
❑ What was schedule compliance?

161

- ❑ Were any of the schedule shortfalls due to incomplete or poor planning?
- ❑ What was the problem?
- ❑ Could it have been avoided?
- ❑ What can we do differently next time?
- ❑ What changes must be made?

Management can assess the quality of planning by periodically raising the questions provided. However, for the process to be successful and meaningful, supervisors must also critique their mechanics during and after each job, as a normal element of on-the-job supervisory responsibilities. Supervisor to planner feedback, and even mechanic to planner feedback, should not wait for the weekly coordination meeting but should be part of the continuous team effort and should be conveyed in a constructive manner at every opportunity. Some operations solicit mechanic feedback by means of a Job Plan Survey (introduced above and appended).

ACTIVITY SAMPLING

The final means of measuring planning quality is through periodic activity sampling. This technique statistically samples the workweek of the maintenance work force to determine the portions spent at **direct work*** (on-site use of tools/wrench time) and that consumed by specific forms of delay (avoidable and unavoidable). It is the avoidable delays that quantify planning shortfalls. This topic is covered in more detail in the introduction.

Other metrics can be found in Chapter 4 - Managing the Planning and Scheduling Function and Chapter 15 - Job Closeout and Follow Up.

---

* This usage of "Direct Work" differs from the accounting term referenced in the calculation of Schedule Effectiveness (Chapter 15).

Computerized Maintenance Management Information System (CMMIS) In Support of Planning, Scheduling, and Coordination

Effective planning, coordination, and scheduling of the maintenance function can be, and for many years was, accomplished without computer support. However, in these days of high technology and rapid, economical data communication, job preparation is accomplished far more efficiently with the support of a sound Computerized Maintenance Information Management System (CMMIS).

Note to maintenance professionals who follow the field: The generally accepted term for maintenance computer systems is Computerized Maintenance Management Systems (CMMS). Computerized Maintenance Management Information Systems (CMMIS) is preferred because current systems by design and by use are not, for the most part, used to manage maintenance but rather to inform about maintenance. Both acronyms are used in this text.

Chapter 17

The "I" is inserted into the CMMIS acronym to emphasize that a computerized support system is only an informational tool and is only one building block of an integrated maintenance excellence process. A CMMIS accomplishes <u>nothing</u> in isolation, but must be integrated with the other twenty building blocks of the "Maintenance Arch" (see Introduction). Bottom-line impact results from actions taken on the basis of information provided by the system, not directly from the system itself.

Fast, flexible access to reliable, current, and comprehensive information is vital if planners and managers are to control the maintenance function on the basis of knowledge rather than intuition. Simply put, it is no longer an economically sound decision to manage a function as critical as maintenance without on-line informational support. Computer support is essential if the full potential of the maintenance control system is to be realized. Only on-line transaction processing systems and networks—people and programs simultaneously retrieving and updating information—satisfy the immediacy required by today's environment.

Integration of the entire Maintenance Arch (Introduction), including the CMMIS, supports and fosters the following:

♦ Efficiency of maintenance resources (both hourly and salaried), thereby lowering unit cost
♦ Improvement of responsiveness and service to internal customers
♦ Improvement of asset reliability, capacity assurance, and equipment up time
♦ Better delivery performance and product quality to external customers
♦ Lower unit costs and increased profitability

Computerization of the work order system allows easier access to large amounts of data enabling analyses too time consuming to perform manually.

A popular phrase regarding many CMMIS's on the market is that they "are not user-friendly." The statement is true. It is also true that the functions and persons to which the systems are least user-friendly are the planner-schedulers.

The maintenance planning system is generally part of a much larger maintenance information system. It is not the intention in this book to discuss all features and characteristics essential to an effective CMMIS, but to concentrate on those capabilities pertinent to work identification, backlog management, job planning, material procurement, logistical coordination, and weekly scheduling. Of course, planner-schedulers are not the only parties interested in maintenance-associated information.

To effectively support the functions discussed throughout this book, the chosen CMMIS (hardware and software) must offer the following characteristics.

The selected CMMIS must be a sound, comprehensive, on-line, real-time, user-friendly, computerized work order control system. If it is not real time, the maintenance staff (planners and clerks) must perform all administrative input and output. A preferable strategy is for all parties to do their own share of informational input and retrieval.

If these responsibilities are not shared, all too often, planners become little more than clerks. It is a sound investment to take a skilled mechanic off the tools to become a planner but it is a poor investment indeed to take a planner off planning and relegate that person to clerical data entry. It is also a problem for skilled mechanics to take excessive time from being on tools to perform data entry. The design of the system should take advantage of all available technology to minimize the time required for input and retrieval. Remember, if a clerk is doing the entry, the mechanic must first write everything down, make sure it is readable, and ensure that the clerk knows what is being referred to or garbage will get into the system.

There is nothing new in this distribution of responsibility. When work order systems first came into use, well over 50 years ago, people requesting maintenance support were expected to submit a written work order. A Work Order Request now only requires filling in requestor-required fields on a form or a computer screen.

Good backlog management features that enable the quantification, by craft and type, of all open work orders, are essential. These features relate essentially to effective coding regarding:

- "Job Status" to facilitate the planner's efforts to keep all work orders moving to completion rather than allowing them to bog down in a state of limbo.

- "Assigned Team" to facilitate the preparation of a weekly schedule fully deploying the resources reporting to each given supervisor.
- "Asset/Equipment" to facilitate the assembly of all ready-to-go backlog that might be performed during access to a given asset.
- "Requestor" to keep internal customers appraised as to the current status of their requests. Ideally, requestors should be able to access this information themselves, on-line.
- "Planner" so that each planner is able to separate his or her work load from the complete backlog.
- "Condition required" because there is regularly a need to separate work that is doable at any time versus that requiring asset down time, of various duration (a few hours, a weekend, periodic programmed access, annual shutdown, etc.)

Features in CMMIS that support effective planning and estimating include:

- Macro-planning to keep resources in balance with the workload. That is, screens to allow calculation of available hours, and deductions for projected PMs and provide an estimate for break down hours. The macro-plan then calculates capacity available for backlog relief. This is called the Work Program process (Chapter 6).
- System capability to store, retrieve, modify, and copy previously developed job plans and estimates from history or planner libraries.
- When planning a job on a specific asset, ready access to related information without backing out of the planning module. Being able to cut and paste is useful. The information in question includes previously cataloged:
  – Job Steps
  – Bills of Material cataloged by machine and major component
  – Current available inventory with capability to reserve same
  – Job Estimates
  – Pertinent safety and environmental procedures, instructions, permits and authorization (hot work, isolation, lockout/tag out, pre- and post- maintenance valve and switch lineups, etc.)

166

- System linkage to current drawings and other reference documents with provision for automatic attachment to planned job packages.

Effective materials management features are essential. Problems in this area are common and constantly threaten mechanic's productivity.

- Cross references
  - Bill Of Materials (BOM) of components by asset.
  - Conversions between manufacturer's part numbers, vendor's part numbers and storeroom item numbers
- Reservation (allocation) of inventory item units to specific planned jobs and release (de-allocation) of same as needed
- Reliable replenishment of authorized stock
- Prompt processing of purchase order requests for direct purchases
- Prompt and reliable notification of receipts

In addition to system capabilities there are practices that make the system work effectively. Below are some effective scheduling procedures:

- Annualized leveling of PM/PdM's with notification as they come due

- Ability to call forward approaching PM's to take advantage of known asset access

- Weekly scheduling for each crew by job, day of the week, and individual to whom job is aligned

- Linkage to project management software

- Good reporting features including Backlog Status, Work Programs, Schedule Compliance, Crew Efficiency, and Age of Reserved Inventory

Planning Major Maintenance Projects

In addition to the many jobs performed by maintenance throughout the year on a day-to-day or
week-to-week basis, maintenance is also respon-
sible for major efforts comprising many tasks. These major tasks include
capital projects as well as major maintenance shutdowns, such as the two
to three week annual turnaround common to many industries. Projects differ in that they require financial justification while the need for maintenance turnaround is evident. This chapter addresses the planning, management and control of these two categories of major effort in a common manner. They are significantly more alike than they are different.

These projects still require effective preparation following the
same process addressed throughout this book. The difference lies in scope
of the work, number of resources required, the limited time frame and
therefore the amount of essential coordination. The magnitude of these
items dictate the application of project management techniques and software which aids thought in terms of networks of simultaneous parallel
activities (arrow, network diagrams) rather than series of sequential, linear
activities. Maintenance planners often use "Microsoft Project", whereas
Project Engineers and Project Managers typically use the more comprehensive "Primavera" software. All project management systems and software applies one of two primary approaches to network analysis. They are
based either on Critical Path Method (CPM) or on Project Evaluation

Review Technique (PERT). CPM terminology is used throughout this chapter.

When these techniques are first deployed within an organization, even the most experienced construction managers are amazed at the opportunities they find to compress and optimize projects similar to those they have been managing for many years. The more important it becomes to complete a project in the shortest period of time, the more essential use of a critical path technique becomes.

Take an annual turnaround as an example. The more days and hours that a plant can be, effectively operated per year, the more profitable the business entity will be assuming the increased throughput can be sold. Within a single month, the annual turnaround typically consumes 35% to 65% of the annual maintenance budget. Obviously, such an event must be budgeted, managed, planned, coordinated, scheduled, executed, and controlled. Critical Path Analysis (CPA) is the means by which major turnarounds and projects are optimized.

Popular thinking is that each turnaround is different. In actuality, turnarounds from year to year are far more alike than different (70% to 80% is the same). If the critical path for common tasks of a turnaround are developed and preserved, the planning for each successive year is greatly facilitated. Effort should then be made to fit the unique activities (jobs) into the available "float" or slack time that exists off the critical path. If all the unique jobs cannot be accomplished without extending project duration more contract resources are added, some unique jobs are excluded from the project, or the turnaround period must be extended.

No attempt will be made here to "train" the reader in application of project management tools. Other writers have already done a better job than we can hope to accomplish within a single chapter. We strongly recommend that you expose your planners to one or more of these works. Once planners are exposed to network analysis, their analytical skills and preparatory thought process grow dramatically.

WHO SHOULD APPLY PROJECT MANAGEMENT TECHNIQUES IN THE MAINTENANCE ARENA

A single designated manager of a given project is essential in any environment. Senior planners usually handle various shutdowns throughout the year and small projects that are expensed.

But the person selected for a capital project or major annual turn-around is usually a manager or an engineer. Either requires support of a Maintenance Planner if internal maintenance resources are to be utilized. Required procedures and proper methods; required parts, materials, tools, and equipment; and access to assets must be established and assembled into planned job packages. These requirements don't come from a project manager or project engineer. They come from a Maintenance Planner-Scheduler.

The requirements of turnaround planning are too significant to load upon a planner already consumed by week-to-week demands. It is generally not practical for the same individual to concurrently plan and schedule normal weekly/daily effort as well as major shutdowns (it is impossible to effectively address both sets of demands concurrently). One essential preparation responsibility or the other is bound to suffer and the planning process will fall significantly short of the essential need.

Therefore, the Turnaround Planner-Scheduler is (usually) not the same person that prepares for the effective execution of maintenance work throughout the year. Planning and Scheduling for major shutdowns is clearly a specialty. Planning for next year's turnaround begins with debriefing this year's turnaround, which is the identification of things that went wrong or could be improved upon. These items are immediately documented to avoid repetition in future years. The intense planning process begins at least four months prior to upcoming turnarounds, and even this amount of lead-time is not sufficient when the sources of major equipment and parts are located in Europe or Asia.

FURTHER DISCUSSION OF THE
CRITICAL PATH TECHNIQUE

Throughout this book, Planning has been defined as "how to" and Scheduling as "when to." CPM techniques contain aspects of both planning and scheduling. Someone still must detail the project to define and estimate the requirements of each activity comprising the network. Developing the network identifies and clarifies the many interfaces and thereby the opportunities to shorten duration of the overall effort. Establishing network interfaces with input of all parties greatly facilitates "coordination"; and fixing earliest and latest start and finish times for each activity facilitates "scheduling."

In deployment of Project Management techniques, the overall effort is divided into activities concluding and events. Completion of all precedent activities is marked by the achievement of an event. By monitoring milestone start times of activities and realization times of events; managers internal to and external to the project team can review progress to ensure compliance with budgets and business objectives.

The process requires the following inputs:

❖ Identification of the individual jobs or activities comprising the overall project
❖ Skills and crew sizes required for each activity
❖ Duration required for each activity
❖ Predecessor and Successor activities for each event (milestones achieved)

Although the network diagram is a better tool for analysis, the "Gannt-bar chart is visibly more readable to many people in the project loop and remains part of the CPM package. The bar chart is also used in resource leveling. The CPM process is to first find the shortest feasible critical path assuming no resource restrictions. The process then seeks to shave peak resource demands by delaying activities that are off the critical path. This can be done only to the limit of their float, so as not to change and lengthen the critical path. The vehicle used to accomplish this is an extrapolation of the bar chart into a resource load chart.

"Float" is the CPM term for slack. This is the duration of delay within each activity that can be absorbed before it becomes critical and extends overall project duration. Three forms of float are quantified to allow project managers and planners to quickly assess the impact of project delays:

❖ Total Float – is the delay in activity start or the increase in activity duration that can be absorbed without delaying project completion.
❖ Free Float – is the delay in activity start or the increase in activity duration that can be absorbed without impinging upon the timely start of any immediate successor activity.
❖ Independent Float – is the delay in activity start or the increase in activity duration that can be absorbed without adversely impacting any other predecessor or successor activity.

CPM is also a budgetary development tool because budgets are associated with each activity and event of the network. It is furthermore a project management and control tool throughout project execution. It provides alarms when:

- ❖ Interim targets are missed (earliest event time on the critical path)
- ❖ Float is reduced (latest event time off the critical path)
- ❖ The rate of financial commitment at any point of the network exceeds projected cash outflow.

Variances between what was planned and what is actually happening must be tracked: are activity start and finish dates being met; are estimated cost working out in reality; are planned resource requirements matching actual; and are deliverables being achieved? If activities are well defined, tracking is not difficult. The key to keeping projects on track is to detect developing problems early enough to take corrective actions. It is not acceptable to be surprised after the situation is already awry.

To effectively use these techniques Project Managers will divide the project into phases. Completion of each phase is marked by the realization of a deliverable. By setting up phase-ending milestones as deliverables, managers outside the immediate project management team project can review progress and ensure compliance with budgets and business objectives.

Steps of the Project Management Process

Phase One – Project Definition

1. Create a vision for the project. Chose the project management team. Establish a physical structure and chain of command to manage the project. Make sure that each member knows his or her role in the overall project. Train the team in project management and related software.

2. Identify and involve the stakeholders to ensure that all essential expectations are considered during project development to avoid scope changes after project initiation.

3. Determine the general scope of work from the engineering schedule and from preventive maintenance inspections. Review all drawings, specifications and other available documentation.

4. Develop a summary statement of the turnaround or project objective and distribute for concurrence ... by signature.

5. Define the deliverables constituting project completion.

Phase Two – Preliminary Engineering

6. Determine pre-shutdown and other preparation. Initiate associated work orders.

7. Determine the extent of partial or complete shutdown of associated processes and adjacent areas.

8. Gather all PM's and backlog jobs that may be candidates to be performed during the shutdown or while project work is being performed in an area. Look forward for PM's that might be accelerated. Reference the previous shutdown in the given area as a starting template, paying close attention to activity sequencing and problems encountered. Do not repeat the same mistakes.

9. Break the project into jobs and individual activities comprising the jobs. There are rules for setting up an activity

♦ An activity has a defined beginning and ending. The better defined, the easier it is to manage the activity.

♦ No subsequent activity can originate from the middle of a prior activity. If one really can, the earlier activity must be split into two activities.

10. Determine general manpower requirements and skill sets, defined by in-house versus contract personnel.

11. As contractors are chosen, check that they can meet their commitments. Contractors go to great length to make a sale. Look at the depth of their organization, the time of year (is it a busy time even without your job). The contractor needs only to win a string of proposals to become over committed. When good contractors are found, the project manager must be a fair customer and avoid negotiating deadlines, milestones and budgets that are unreasonable. Reliable contractors should be regarded as and treated as partners.

12. List all heavy or specialized equipment likely to be required, such as cranes, forklifts, scaffolding, compressors, welding machines, or torque wrenches, and assure their timely availablility.

13. Estimate elapsed time needed for each activity. Depending on the risk associated with schedule delinquency, three estimates may be made (optimistic, probable and pessimistic). Studies show that the more pessimistic estimates best reflect actual. This may be so in the first instance of a given turnaround. But as planning improves the "probable" should become reliable. Most overruns occur as a result of preparatory shortfalls.

14. Identify prerequisites for each activity. Determine what activities must be completed before each activity can be started. Block walls cannot be laid-up until footers are cured. A particular activity may be dependent upon completion of several other activities.

15. Post the required data to the CPM software. From this information the software builds a "model of the project" in the form of a critical path network with time-line bar chart, resource demand chart, and projected cash flow. Determine how long the project will take. The activities constituting the longest path through the project are the critical path. Slippage in any critical path activity will result in the project being completed late.

For many activities, late starts will not affect the overall project completion date. Time before a delayed activity becomes critical is called "float". Until float is consumed, the late activity is not critical. When float runs out, non-critical path activities become critical and alter the critical path.

This is the opportunity to work the project through, on paper. Assuring that the movement of materials, people and equipment is safe and reasonably efficient. Find the shortfalls in this paper exercise, not during live execution.

The first planning pass is done in relative time (elapsed days from day one). When ready to plug in an actual start date, the software will automatically post dates to the time-lines.

16. From labor estimates, contractor requirements, material and parts estimates, and required equipment establish a comprehensive budget for the overall project.

Phase Three – Justification and Funding

17. When the first cut of the project is established, risks are identified and quantified for inclusion in the forthcoming request for authorization and funding.

18. In the case of capital projects, a return on investment is established and added to the risk analysis. They are jointly presented for comparison to ROI hurdles prior to authorization.

Phase Four – Detailed Project Planning

19. Decide on need and status of materials, such as valves, internals, etc., and make sure that the parts and materials will arrive in sufficient time for checkout prior to use.

20. When the workload has been established, define supervisory needs and responsibilities.

21. Identify other staff support requirements such as clerical, data entry and timekeeping.

22. Initiate P.O.'s and WO's for contractors, rental equipment, etc.

23. Arrange for temporary office space with required support equipment (phones, copy machine, etc.).

24. Make transportation arrangements for supervisors and crews.

25. Arrange for material and tool trailers if needed.

26. Make arrangements with Production for temporary storage (lay-down space) of large equipment sections temporarily removed in the turnaround process.

27. Arrange for dumpsters for collection of waste material.

28. Order portable toilets if needed.

29. Secure a list of all contract workers and arrange for their safety orientation.

30. Assemble drawings, wiring diagrams, shop drawings and other reference documents. Add them to appropriate Planned Job Packages.

31. Secure permits covering safety, fire, and regulatory requirements (local, state and federal).

32. Summarize the project responsibilities of each individual supervisor. Provide them with the associated Planned Job Packages. The package should include:

◆ Turnaround objectives
◆ Turnaround schedule
◆ Detailed Work Orders for all jobs to be supervised
◆ Copy of the turnaround organizational chart
◆ List of responsibilities by supervisor

176

- Progress report forms with instructions for their use
- Instructions regarding contractor daily time reporting
- Set of craft work rules
- Telephone and beeper list
- List of helpful reminders

33. Provide a list of telephone and beeper numbers to all personnel key to the project.
34. Go over this list again to verify that preparation is complete.
35. Distribute the final project schedule to all appropriate parties.

Phase Five – Project Execution

36. Begin to receive and stage parts, materials and equipment.
37. Execution of the project or turnaround is now ready to begin.
38. Continuously monitor and post progress relative to plan. This enables the software to provide real-time alerts when activities fall behind or expenditures exceed budget. Identify slippage promptly and take corrective action. Expedite as necessary. Add resources (overtime or additional contract support) to overcome the shortfalls in order to return to plan. Focus on and protect the critical path, but keep an eye on the other paths. Make sure they do not exceed their available float.
39. Change Orders – Scope creep should be resisted. Some changes won't go away, but the overall project must be protected. Protect the boss from adverse surprises. Protect the team from disruptions that changes create. Manage the changes A change order process should be created at the beginning of the project. How many times has it been said, "While we are doing this, it would be easy to also do this other thing"? If change is accepted, everyone up the ladder should sign off on it and accept the consequences in terms of extended duration and increased cost.

Phase Six – Project Completion and Close-Out

40. Complete cleanup of the site
41. Punch list completion
42. Turn-over, quality assurance, life safety testing
43. Start-up

44. Delivery of "As Installed" drawings and other contracted deliverables (recommended PM routines, BOM and recommended spare parts

45. End-user acceptance

46. Post completion review. This is a debriefing to establish lessons learned in order preclude their repetition.

47. Project Closeout Report should be written as soon after completion as possible. It documents what happened, what didn't happen, and how problems were fixed. It contains a set of project documents.

Phase Seven – Project Review (six months after completion)

48. Actual financials realized should be compared to projection. Failure to deliver expectations damages credibility, which appropriately influences approval of future capital requests.

Table of Appendices

APPENDIX A

JOB DESCRIPTION

Title: Maintenance Planner/Scheduler

The job of Planner/Scheduler is critical to effective maintenance. It is essential that the positions be staffed by personnel with required aptitudes and temperament. Planners should be assigned 100% to their function. Planners can be used to backup maintenance supervisors when absent for one reason or another. However, someone in turn must backup the planner. Leaving the planner role uncovered must not be condoned. It conveys lack of commitment to the function and implies the function is not essential (and in the next downsizing can be done without).

Reports to: Maintenance Manager

Position Scope: The primary role of the maintenance planner is to improve work force productivity and work quality by anticipating and eliminating potential delays through planning and coordination of labor, parts and material, tools and equipment, permissions, specialized documentation and equipment access.

The planner/scheduler is responsible for the planning and scheduling of all maintenance work performed in the area to which he/she is assigned. He/she maintains liaison and coordination between the production and maintenance organizations; maintains appropriate records and files to permit meaningful analysis and reporting of results or work done.

Responsibilities and Duties: In the performance of his/her duties, the planner/scheduler:

1. Is the principal contact and liaison person between the maintenance department and the plant departments served by maintenance and ensures that the

production areas or functions to which he/she is assigned receive prompt, efficient and quality service from the maintenance function and ensures the maintenance function is given the opportunity to provide this service.

2. Is responsible for long-range as well as short-range planning. Long-range planning involves the regular analysis of backlog relative to available resources. These two basic variables must be kept in balance if a proactive maintenance environment is to be established and sustained.

3. The planner receives all work orders from the requesting departments of the areas to which he/she is assigned, except for emergency work that is requested of the appropriate maintenance supervisor for immediate attention.

4. Makes any additional sketches diagrams, etc., necessary to clarify the intent of the work order.

5. Reviews and screens each work order to see that is has been properly filled out:
   — Requested Work is clearly described
   — Check if the priority and requested completion date are realistic and provide practical lead-time
   — Charge numbers and other coding are complete and accurate
   — Authorization is proper
   — Discusses the details with the originating department as appropriate

6. Assures that work requested is needed. If need is questioned and not readily resolved with production or requesting personnel, refers the work order to the maintenance manager.

7. Reviews with engineering those work orders requiring engineering design.

8. Maintenance work orders which can be planned out but which require participation by shop or other crews are copied (cross-order) and provided to the appropriate planner for planning of the supplemental work.

9. Examines jobs to be done and determines scope and best way to accomplish the work. Consults with requester or maintenance supervisor when necessary.

10. Obtains blueprints, drawings, instructional manuals and special procedures, as needed, from files or other sources. Makes any additional sketches, diagrams, etc., necessary to clarify the intent of the work order.

11. Identifies and obtains (requisitions, orders, kits as appropriate and in keeping with procedures) determinable materials, entering material needs on the work order. Determines if critical items are in stock by verifying availability with stores and reserving same.

12. Ensures the safety needs are given a top priority in work planning and

scheduling.

13. Estimates jobs showing sequence of steps, the number of mechanics and required man-hours for each step.

14. Estimates cost of each work order in terms of direct labor, materials required and total cost.

15. Maintains backlog files of work orders awaiting scheduling in accordance with their priority and requested completion date. Those jobs unplanned, requiring engineering, waiting for materials, waiting for down-time, etc., are filed accordingly. When ready for scheduling, work orders are filed by supervisor by required completion day. Ideally, this filing is accomplished within a computerized system.

16. Once a job is planned and estimated, prior to scheduling, verifies the availability of parts, materials and special tools required for its execution.

17. Planner should have complete knowledge of each department's PM workload in order to better schedule work orders in that area.

18. Reviews schedule status and forecast of manpower availability on a regular basis.

19. Develops a maintenance work schedule for the maintenance area supervisor. From the backlog files for each crew, selects a group of work orders with manpower requirements matching the capability of the identified work forces, taking into account any work carry over from jobs previously scheduled. Identifies any special skill requirements and makes such arrangements with the responsible planners.

20. Allocates manpower and coordinates these requirements through maintenance supervisors and the maintenance manager.

21. In the selection of jobs for scheduling, meeting the deadlines established by the requesting department and maintaining preventive maintenance schedules is essential. If any work orders cannot be scheduled within the time expectation, the requesting department management and requester are promptly notified so that appropriate action can be taken to get the work done in a satisfactory and timely manner.

22. Attends meetings with the production planning department and participates in the overall plant scheduling of the following week's work, and negotiates for downtime "windows" during which preventive and corrective maintenance requiring downtime can be performed. Finalizes own schedules for which he/she is responsible, ensuring that the work scheduled balances with the man-hours available so that a full day's work is provided each person.

23. Recommends equipment to be included in preventive maintenance programs.

Appendix

24. Schedules preventive maintenance and other planned work in coordination with production and maintenance supervisors.
25. On the basis of firm work schedules, coordinates requisition of all predetermined parts, materials and special tools and ensures that equipment to be worked on will be available and ready. Arranges for any safety inspection, fire and standby watch.
26. Issues approved schedules together with relevant work orders and other planning documents to area supervisors. Discusses "planning packages" as necessary with special instructions or considerations to be observed in the execution of the jobs and reviews new jobs coming up in the future. (All work orders except emergencies come through the planner.)
27. Follows up to ensure the completed schedules and work orders are returned at the proper time. Carefully reviews completed schedules and corresponding work orders returnrd by the maintenance area supervisors monitors work order progress, and prepares associated reports .
28. In accordance with standard practice, provides the clerks with all documents for reporting or closeout.
29. Remains abreast of accumulated cost to standing work orders versus budget.
30. Promotes the conservation of energy.
31. Maintains close contact with other planners to ensure coordination of complex multi-skill field and shop jobs.
32. Schedules weekly meetings with production and maintenance supervisors concerned with the areas for which he/she is responsible, consulting them regarding facilities or equipment to be maintained. Makes recommendations to production concerning long-range maintenance needs and, in collaboration with production, prepares a weekly forecast of all jobs expected to be scheduled for the following week.
33. Maintains a list of planned work orders requiring equipment to be down, so that some or all can be performed in the event of an unscheduled. These lists, by asset, are reviewed and updated weekly.
34. Develops a file of standard work orders (plans) for regularly recurring repair jobs, based on historical experience, to simplify the planning process.
35. Reviews with the maintenance area supervisors the actual labor expended versus estimated labor and material used for completed jobs, in order to determine corrective measures needed to improve the accuracy of estimating and improving methods of doing work.
36. Job Estimates are continually refined and thereby reflect improving accuracy and consistency; representing fair as well as challenging expectancies.

37. Assists maintenance and production management in periodically analyzing costs and, where necessary, recommends corrective action needed to reduce maintenance costs.
38. Keeps the maintenance manager properly informed on all abnormal or critical situations and seeks advice on matters outside of the planner's knowledge or authority.
39. Make recommendations for system improvement.
40. Maintains necessary records and files and prepares and distributes meaningful and accurate control reports.
41. Performs other tasks and special assignments as requested by the maintenance manager.
42. Develops and maintains planner reference systems (library) including a file of Planned Job Packages for recurring jobs, plus labor and material libraries for each equipment center.
43. Recommend additions to authorized stock.

Measurement of Position Performance

The planner/scheduler is measured in the performance of his/her duties and execution of assigned responsibilities by:

1. Labor effectiveness of the area to which assigned as measured by work sampling.
2. Orderly placement of work orders on the schedule(s) and their completion within the specified time frame.
3. Jobs worked and completed per schedule.
4. Improvement in mean time between equipment failures.
5. The accuracy of estimates of labor and material.
6. The improved use of labor and improved plant condition as expressed by a reduction in emergencies, overtime hours worked and in unscheduled labor worked, and reduction of contractor support.
7. The timely and accurate preparation and distribution of meaningful control reports.

Position Goals

1. To ensure that production areas or functions that he/she serves receive prompt, efficient and quality service from the maintenance function

enabling them to operate at a high level of efficiency.

2. To ensure the maintenance function is given the opportunity to provide production with the service it requires.
3. Accurately define and estimate work requests.
4. Properly prepare and distribute meaningful control reports.
5. To meet the needs of customers, both internally and externally. Our future is in the hands of our customers. We will be courteous, helpful, and responsive to them always. We want our people to be focused on our/their customers (internal and external).

Relationships

1. Reports to the maintenance manager.
2. Works closely with production supervision.
3. Works closely with maintenance supervisors.
4. Works closely with stores and purchasing personnel.
5. Maintains good working relationships with other organizational units in the plant.

Requirements, Qualifications and Selection Criteria

1. Mechanical/electrical background necessary and tech school background desired.
2. Adequate craft knowledge to estimate labor hours and materials and to visualize the job to be performed.
3. Good oral/written communication skills and possession of tact.
4. Good administrative and mathematical skills with willingness to handle paperwork
5. Have or able to acquire a working knowledge of personal computers in a reasonable training period (typing skills helpful)
6. Good planning and organizational skills.
7. Ability to understand what constitutes good instructions
8. Able to read blueprints and shop drawings
9. Sketching ability
10. Understanding of the proper use of work orders, priorities, scheduling, etc.
11. Ability to keep multiple jobs in controlled motion—simultaneously.
12. Ability to bring about order—from chaos.
13. Orientation and commitment to customer service.

14. Style and capability commanding respect within both maintenance and operating organizations

Exclusions from Planner responsibilities are:

- Involvement in daily emergency and urgent requests. There is no opportunity to plan such work. If any planning is to be accomplished, the Planner must focus on tomorrow and beyond.
- Daily assignment of individual mechanics to specific jobs.However, the Planner should develop the recommended manpower deployment plan for assignment of individuals to specific jobs—this is a preparatory exercise.
- Maintenance Engineering including development of the PM/PdM system, analysis of Equipment History and re-engineering points of repetitive equipment failure.

Appendix B

Position Description

Title: Manager of Maintenance Support Services
Often the Maintenance Manager has a line versus staff orientation. Very likely he/she was once a Master Mechanic. His/her strength is still at the job site. Others are more oriented to work preparation and administration. Small organizations cannot justify a position to mange the support staff. Larger organizations can.

Reports To: Maintenance Manager

Position Scope: Planning and Scheduling, Maintenance Engineering, Administrative Support, and Maintenance Purchasing & Stores (if those functions are in the maintenance organization).

Responsibilities and Duties: The Manager of Maintenance Support Services is the primary:

- Champion and Chief Advocate of the Maintenance Excellence process

- Marketer and Seller of Maintenance Control throughout the organization

- As manager of all maintenance support services (staff not line), he or she is the Chief Planner/Scheduler, chief of maintenance engineering, CMMS administrator and head librarian for maintenance and planning technical libraries

- Owner and Controller of the CMMS and as such:
  - Controls system security
  - Steers and continuously seeks system improvements (coding, user friendliness, application and reporting)
  - Trainer of all parties regarding their responsibilities to the system and their usage thereof
  - Enforcer of proper system input and usage by confrontation of the abusers

187

Appendix

- Leader of Data Base, Job Plan, and Job Estimate refinement

- Quality Assurance & Control leader of the CMMIS, Planning, and Scheduling, and RCM (PPM and Root Cause Analysis) processes

- Analyzer of system information and trends of improvement and maintenance metrics ... including the balance of resources with workload

- Develops and carries forth the consensus issues and recommendations of the Maintenance Support Services Team

- May be responsible for the maintenance store room and for purchasing

- Executes special studies requested by management or self generated

APPENDIX C

TYPICAL SOURCES OF PLANNED MAINTENANCE WORK

Sources of Planned Work	Percent of Total Planned Work
Results of PM inspections	30%
Scheduled component replacements	20%
Overhauls/rebuilds	15%
Internal Customer Input (Operators and Supervisors)	10%
Engineering project support	8%
Safety work	5%
Analysis of repair history	5%
Management directed work	4%
Service requests	2%
Accident damage	1%
Total Planned Work	100%

APPENDIX D
PLANNER TIME LOG

Name: _____   Date: _____

Time	Act. Code	Time	Act. Code	Time	Act. Code	Time	Act Code	Time	Act. Code
5:30		8:05		10:40		1:15		3:50	
5:35		8:10		10:45		1:20		3:55	
5:40		8:15		10:50		1:25		4:00	
5:45		8:20		10:55		1:30		4:05	
5:50		8:25		11:00		1:35		4:10	
5:55		8:30		11:05		1:40		4:15	
6:00		8:35		11:10		1:45		4:20	
6:05		8:40		11:15		1:50		4:25	
6:10		8:45		11:20		1:55		4:30	
6:15		8:50		11:25		2:00		4:35	
6:20		8:55		11:30		2:05		4:40	
6:25		9:00		11:35		2:10		4:45	
6:30		9:05		11:40		2:15		4:50	
6:35		9:10		11:45		2:20		4:55	
6:40		9:15		11:50		2:25		5:00	
6:45		9:20		11:55		2:30		5:05	
6:50		9:25		12:00		2:35		5:10	
6:55		9:30		12:05		2:40		5:15	
7:00		9:35		12:10		2:45		5:20	
7:05		9:40		12:15		2:50		5:25	
7:10		9:45		12:20		2:55		5:30	
7:15		9:50		12:25		3:00		5:35	
7:20		9:55		12:30		3:05		5:40	
7:25		10:00		12:35		3:10		5:45	
7:30		10:05		12:40		3:15		5:50	
7:35		10:10		12:45		3:20		5:55	
7:40		10:15		12:50		3:25		6:00	
7:45		10:20		12:55		3:30		6:05	
7:50		10:25		1:00		3:35		6:10	
7:55		10:30		1:05		3:40		6:15	
8:00		10:35		1:10		3:45		6:20	

Activity Codes: DS = Daily Scheduling ME = Material Expediting MT = Meetings JP = Job Planning
WP = Weekly Planning EN = Engineering PR = Problem Resolution
MO = Material Ordering CO = Counseling FO = Schedule Follow-up
AA = Avoidable Activity UD = Unavoidable Delay

APPENDIX E
DETERMINATION WORKSHEET
RATIO OF CRAFTSMEN TO PLANNERS

Points Assigned

Planning and Scheduling Structure:
- ➢Separate from material coordinating (vertical structure) - 1 point
- ➢Combined (horizontal structure) - 2 points   ( )

Number of Crafts Coordinated
- ➢One - 1 point
- ➢Two - 2 points
- ➢Three - 3 points
- ➢Four - 4 points   ( )

Level of Planning
- ➢Craft and general description with schedule - 1 point
- ➢Craft, general instructions, special tools and major materials with schedule - 3 points
- ➢Craft, specific instructions, tools, materials, prints and schedule - 5 points
- ➢All the above plus work methods described - 7 points   ( )

Level of Estimating
- ➢Estimates or historical averages - 1 point
- ➢Slotting against benchmarks or labor library - 3 points
- ➢Analytical estimating - 5 points
- ➢Engineered Standards – 7 points   ( )

Inappropriate Responsibilities (additive)
Sourcing – 1 point
Procuring – 1 point
Expediting – 1 point
Receiving – 1 point
Stocking – 1 point
Picking and kiting – 1 point
Staging and securing – 1 point
Delivery to scheduled job site – 1 point   ( )

Total Points: _____

APPENDIX E continued

Conversion Table

Total Points	Craftsmen:Planner Ratio
4 to 7	30:1
8 to 12	25:1
13 to 17	20:1
18 to 22	15:1
23 to 26	12:1
27 to 30	10:1

APPENDIX F
Labor / Material Library

Class 08 Type 01			Desc: GROTNES RIM ROLL 600			
W/O	Div	Line		Type Work		Status
Job						

Seq No	Task Sequence		MR M/HR		TK M/HR		PF M/HR		EL M/HR		SF M/HR		M/HR
0.10	Remove Top Hydromotor	2	2.00										
0.20	Remove Top Roll Shaft	2	5.00	1	0.50								
0.30	Remove Top Front Trunion	2	2.00										
0.40	Remove Lower Hydromotor	2	2.00			1	0.50						
0.50	Remove Lower Roll Shaft	2	5.00	1	0.50								
0.60	Remove Front Trunion	2	2.00										
0.70	Remove Flotork	2	1.00			1	0.50						
0.80	Remove Main Hyd Pump	1	1.00			1	0.50						
0.90	Remove Small Hyd Pump	1	1.00			1	0.50						
100	Remove Main Drive Motor	2	3.00	1	1.00			1	0.80				
110	Remove Rim Loader	2	2.00			1	0.80	1	0.50				
120	Change Bearings in Crank	2	16.00										
130	Remove Loader Control Arm	1	1.50										
140	Remove Small Drive Motor	2	2.00					1	0.50				
150	Remove Pitman Arm	1	0.60										
160	Remove Guide Roll Assem.	2	3.00			1	0.50						
170	Replace Guide Roll Assem.	2	3.00			1	0.50						
180	Replace Pitman Arm	1	0.80										
190	Replace Small Drive Motor	2	3.00					1	1.00				
200	Replace Loader Control Arm	1	1.50										
210	Replace Rim Loader	2	4.00			1	0.70	1	0.80				
220	Replace Main Drive Motor	2	3.00	1	1.00			1	1.00				
230	Replace Small Hyd Pump	1	1.00			1	0.70						
240	Replace Main Hyd Pump	2	1.50			1	0.70						
250	Replace Rotac	2	1.50			1	1.00						
260	Replace Ft Lower Trunion	2	2.00										
270	Replace Lower Roll Shaft	2	6.00	1	0.60								
280	Replace Lower Hydromotor	2	2.00			1	0.50						
290	Replace Top Ft Trunion	2	2.00										
300	Replace Top Roll Shaft	2	6.00	1	0.50								
310	Replace Top Hydromotor	2	2.00			1	0.50						

SEQ			Material Estimates		
0.1001		0.001	Vickers Hydromotor	45M110A-11C-10	
0.2001	334663	0.001	GR-44 Roll Shaft	SAME	
0.2002	509055	0.002	Torrington Roller Bearing	HJ-12415448	
0.2003	334616	0.002	Grotnes Race	A-3508-107	
0.2004	508604	0.002	Timken Cone	HM-926747	
0.2005	508602	0.002	Timken Cup	HM-926710	
0.2006	507506	0.004	Garlocx Klozure Seal	53X3548	
0.2007	334646	0.001	Sleeve	A-3508-106	
0.2008	508045	0.001	Timken Lock Nut	TAN-124	
0.2009	508043	0.001	Timken Lock Washer	TW-24	
0.201	525027	0.001	Sier Bath Cpling 2 In Bore	1/2 In Key	
0.3001		0.001	Front Upper Trunion	C-3508-120	
0.3002		0.002	Trunion Sleeve	A-3508-170	
0.4001		0.001	Vickers Hydromotor	45M110A-11C-10	

Note: These estimates account only for Direct Work

Appendix G

Recognizing the Pitfalls in Planning for Others

A. Problems

A.1 Competitive Feelings
Two or three groups plan maintenance jobs (supervisors, technicians and planners). Each thinks their own efforts are the best. The thoughts go as follows: If we execute their plans well, they gain/we lose. If they look good, we look bad by comparison. Such feelings may produce reactions that outright sabotage planner efforts and the transition from a reactive to proactive environment.

A.2 Communication Problems
It is believed there is more room for error, distortion, and misdirection when plans formulated by one group need to be transmitted in their entirety to another group for execution.

A.3 Less Understanding
Doers report greater difficulty in understanding the plans that others assigned to them than with their own plans.

A.4 Less Flexibility
A high percentage of doers think that their own plans are more flexible than plans assigned to them.

A.5 Poor Use of Available Manpower
Doers think that their own plans make better utilization of available man-power.

A.6 Less Commitment
Some believe that persons not involved in planning for their own jobs are less committed to seeing that job planning yields positive results.

A.7 Less Effort to Make the Plan Work
Most maintenance technicians believe they can best plan their own work. This conviction is acted out by expending great effort to ensure that their belief is validated by unfavorable outcome of planner efforts.

A.8 Less Sense of Accomplishment
There is less sense of achievement when others define job plans.

Appendix

B. Possible Remedial Actions

B.1 Minimize Competition

Planners brief supervisors, supervisors brief mechanics, make the assignment and convey planned job packages to the mechanics. Supervisors and planners must avoid creating a situation in which technicians see themselves in a zero-sum game. If the plans succeed, the doers must not lose status, prestige, power, or material benefits. They must experience the benefits of effective planning. Conditions must be established in which supervisors, technicians and planners share common goals. The division of labor should be seen as benefiting each of the three partners.

B.2 Improve Communications

Maintenance technicians must have and believe they have ample opportunity to question supervisors for clarification of plans. Supervisors and planners need to assess whether or not they are overestimating or underestimating the ability of others to comprehend plans and judge whether they are transmitting too much or too little, with or without adequate documentation and visuals.

B.3 Promote Understanding

Understanding of the plan can be fostered by ensuring that the plan itself has been created in a way to minimize its ambiguities. Repeating instructions may increase reliability of understanding. Ask the receiver to paraphrase the instructions back. Promote feedback. Use the Planned Job Survey (presented here in the appendix). Instructions need to be simple enough to be understood by the least capable doer. If he can understand the stages of the plan, all others can also understand it.

B.4 Explain the Importance

Tactical decisions must be made based upon conditions encountered. However, those decisions are to be made by the supervisor. Technicians should be encouraged to recommended essential changes to the supervisor, but not to make them unilaterally. The safety, environmental, and customer communication necessities of this view must be communicated for understanding.

B.5 Maximize Effective Use of Available Manpower

Utilize the scheduling format presented in Chapter 14. There is a thorough approach to scheduling.

B.6 Promote Commitment

Doers can be consulted at various stages in the planning process. Wherever possible, the ideas of the doers can be incorporated in the plans; or when such ideas are unusable, the reasons can be discussed with doers. Also their feedback is reflected in refined job packages, thereby increasing ownership.

Appendix

B.7 Promote Confirmatory Behavior

Planners and supervisors need to communicate confidence in accuracy and expectation of plans together with reasons for their confidence, while technician reservations must be openly discussed and resolved. Although the plans may be sound, if the technicians are left to believe the plans are unrealistic, they may behave in a way to confirm their own doubts. Proper supervision should assure that no negative actions are allowed to persist.

B.8 Foster a Sense of Accomplishment

Plot, post and communicate the trends against baselines and goals. Recognize the small improvements. It won't take long before they add up to major improvement. Give credit to the technicians. Performance of individual members working as a group improves the most when they receive constructive information about their individual efforts as well as the group's success as a whole, particularly if the expectations are challenging. Equally useful is meaningful feedback by one member of the team that will improve the future effectiveness of the entire team.

APPENDIX H   ASSESSMENT OF CURRENT STATE

This series of appendices (G through M) is extracted from a broad Maintenance Assessment Process covering all twenty-one essential building blocks depicted as "The Maintenance Arch" (Introduction). The complete process is available through the authors. Appendix section M has scoring information from the authors database.

Assessment : Organization Structure and Size

Success of maintenance operations begins with a sound and efficient organizational structure, properly designed and adequately staffed with competent personnel working together. The organization may be designed to reflect management's attitudes and objectives, or it can be allowed to evolve in response to requirements. The former approach is more readily controlled and can generally be expected to yield better results.

The Maintenance organization should be structured for responsiveness to facility needs while ensuring intrinsic functional effectiveness through the systemic use of labor, material, equipment, capital, and technological resources.

Response Statements	Numerical Assessment
	1ST 2ND

1. A current and complete organization chart is available and posted showing every position from the Maintenance Manager down, with reporting relationships, and includes supervisors, engineers, planners, clerks, mechanics, janitors, stock and tool attendants, guards, etc.

    The table of organization is clear as to who is responsible for which crafts or areas of plant and all personnel know where they fit within the organization. Periodic reviews are made to determine if structure is still appropriate considering departmental obligations and objectives.

    – Current, complete, effective and posted. (2 points)
    – Available but lacking or not posted information. (1 point)
    – Not available. (0 points)   (   ) (   )

2. Maintenance is an excellent career opportunity. It carries status, recognition and rewards sufficient to attract high caliber, young 'people on a career basis:

    – Salaries, wages, and other rewards are sufficient to attract, retain, and motivate competent maintenance personnel (salaried and hourly).(1 point)   (   ) (   )

      – Career opportunities elsewhere in the organization are not
        blocked by extended maintenance assignment. (1 point)   (   ) (   )

3. The organizational concept allows Maintenance personnel to f
    unction as efficiently as possible:

   •   There is a clear-cut separation between line and staff. The three (3)
        dis-tinct functions within maintenance responsibilities (work execution,
        planning and scheduling, and maintenance engineering) are
        recognized, segregated yet coordinated. As long as these duties are
        delegated, effectively performed and posted, full credit is given
        even if the functions are not performed by different people.

        – Both planning and engineering separate from line (3 points)
        – Planning separate (2 points)
        – Engineering separate (1 point)
        – Line only (0 points)   (   ) (   )
        – The line structure is responsive to the triple nature
            of maintenance work (regular and periodic routine
            responsibilities, response to emergency and urgent
            demands, and timely backlog relief). Similar and
            closely-related activities are collected into logical
            groups that facilitate the efficient execution of
            required work.   (   ) (   )
        – PPM routines occur on schedule. (0-2 points)   (   ) (   )
        – Legitimate emergencies are responded to promptly.
            (0-2 points)   (   ) (   )
        – Backlog jobs are completed within requested timing.
            (0-2 points)   (   ) (   )
   • There is a realistic balance between popular concepts
      of participative management and traditional controls
      structure. (0-2 points)   (   ) (   )

   • The line structure balances resource decentralization and
      centralization. To minimize response time and maximize
      equipment availability, areacrews are decentralized as far as
      practical, but functional shop crewsare centralized to assure
      effective utilization of all resources, and timely backlog relief.
      Small plants normally do not fragment crews, therefore,all
      points are awarded if crews are centralized. (1 point)   (   ) (   )
   • Maintenance responsiveness can be characterized as:
      – Always prompt to job; immediately responds after one call.
        (4 points)
      – Usually responds immediately. (3 points)
      – Usually responds within a reasonable amount of time. (2 points)

Appendix

- Sometimes prompt; sometimes two calls required. (1 point)
- Never prompt despite repeated requests. (0 points)   ( ) ( )
- The concept of process teams with mutual support between
  Maintenance and Production personnel is part of organizational philosophy and direction. This is the future of
  operating/maintenance organizational structure, particularly
  for heavily automated process industries. (1 point)   ( ) ( )
- The number of organizational levels is at the necessary
  minimum. Decision can be made and carried out at the
  lowest appropriate level. Communication and team work
  are facilitated. (1 point)   ( ) ( )
- "High-Involvement" work methods (Quality Circles,
  Work Teams or Self-Management Groups) are evident
  throughout Maintenance. (1 point)   ( ) ( )

4. Supervisors are not so overloaded with administrative or other
   duties that performance of their direct supervisory duties is
   hampered. (1 point)   ( ) ( )

5. Maintenance planning and scheduling is a formal and disciplined
   operation with one person in charge and with sufficient capacity;
   approximately one Planner/Scheduler for each twenty mechanic
   positions authorized. (0-2 points)   ( ) ( )

6. Coordination of outside contractors is also adequately provided
   for. Each million dollars of such contracts, on an ongoing annual
   basis, warrants an additional position (planner, contract administrator
   of project engineer). (1 point)   ( ) ( )

7. Sufficient maintenance engineering capacity exists, with
   approximately one maintenance engineer position (not necessarily
   degreed) for every forty craft positions. The primarily mission of the
   maintenance engineer is to determine causes of failures and recommend corrective action. Only by engineering failures out of equipment can the maintenance function truly fulfill its mission and
   contribute toward improving profitability of the company. If enough
   maintenance engineering capacity exists, either primary or collateral
   duty, and the function is effectively performed then full credit will be
   given. (0-2 points)   ( ) ( )

8. Sufficient clerical support exists, with one clerk for every forty
   wage positions. This allows planners to spend their time planning,
   and supervisors supervising. Clerks are effective from the standpoint
   of cost, dexterity and aptitude. (1 point)   ( ) ( )

Appendix

9. Supervisory span of control is within manageable limits; one direct supervisor for eight to twelve mechanics:
   - Between 5 and 12 to 1 (2 points)
   - Between 13 and 15   (1 point)
   - Above 15 or below 5 (0 points)   ( ) ( )

10. Hourly workers are adequately supervised on dark and weekend shifts.   (1 point)   ( ) ( )

11. Supervisory effectiveness is not diluted by policies or practices preventing direct instruction from foremen to workers. (1 point)   ( ) ( )

12. An Area Supervisor plan is in effect and supervisors are given across-the-board responsibility in their areas regardless of the organizational source of needed skills or capacity. (1 point)   ( ) ( )

13. Each supervisor possesses a copy of his own and his crew's job descriptions (up-to-date) as well as those of his immediate manager and others with whom he has close working contact:
   - All of the above (2 points)
   - Several of the above (1 point)
   - None of the above (0 points)   ( ) ( )

14. Positions with incumbents retiring within two years have been identified and potential replacements are undergoing appropriate training or an appropriate plan of action is in progress to prevent deterioration of mechanic skills and responsiveness. (1 point)   ( ) ( )

15. The number of recognized crafts and jurisdictional agreements are suitable to plant needs. Non-union plants, omit jurisdictional agreements. (1 point)   ( ) ( )

16. There are sufficient cross-trained mechanics and general helpers to balance craft workloads. (1 point)   ( ) ( )

17. Routine and urgent repair work is adequately controlled:
   - Area assigned mechanics are not excessive. If no area assigned mechanics, full credit is given. (1 point)   ( ) ( )
   - Fixed routine assignments charging to "standing" work orders are not excessive. (1 point)   ( ) ( )
   - Personnel allocated to emergency/urgent response do not exceed 20% of the total maintenance work force.(1 point)   ( ) ( )

   • Maintenance forces assigned to shifts without supervisors or lead men are sized to cover only emergency/urgent or

routine work which is best performed by unsupervised shifts
due to better equipment. (1 point)   ( ) ( )

18. The Maintenance organization structure is supportive of daily
operation, internal as well as external to the department.
   – Communication channels and procedures are designed for
   clear and quick transmission of information. (1 point)   ( ) ( )
   – Relationships with staff departments (such as Purchasing,
   Stores, Plant Engineering, and Accounting) are clearly
   defined. (1 point)   ( ) ( )
   – Changes in systems, procedures, and people relations are
   readily assimilated. (1 point)   ( ) ( )

19. Operating departments have designated a specific individual as
their maintenance liaison. (1 point)   ( ) ( )

20. There is an effective training program for Maintenance Planners:
   • Formal classroom training (1 point)   ( ) ( )
   • On-the-job training (1 point)   ( ) ( )
   • Refresher training (1 point)   ( ) ( )

21. It encompasses:

   • The Planner's role within the overall maintenance management
   program. (1 point)   ( ) ( )
   • Job breakdown and analysis. (1 point)   ( ) ( )
   • Materials Requirements determination. (1 point)   ( ) ( )
   • Job estimating. (1 point)   ( ) ( )
   • Weekly scheduling detailed by man, by job, by day.(1 point)   ( ) ( )
   • Communication, coordination and follow-up. (1 point)   ( ) ( )
   • Expediting. (1 point)   ( ) ( )
   • Planning for major shutdowns, turnarounds, outages.(1 point)   ( ) ( )
   • Computer and/or CMMIS training (1 point)   ( ) ( )

Element Summary:

Score: Points Awarded / Potential Points
÷ Potential Points   ( ) ( )
                                                         60   60

= Current State of Effectiveness-Organization   0.   0.
   (Carry to 3 places)

Points Lost: 60 Potential Points – _____Points Awarded =_____

Appendix I

Assessment: Computer Information and Support

Fast, flexible access to current information is vital if managers are to effectively control the Maintenance function. Good manual control systems have partially fulfilled the informational need for years, but computer support is essential if the full potential of the Maintenance Control System is to be realized. Batch processing is no longer good enough. Major decisions hinge on the availability of current information.

Reliable computer power is essential in order to provide immediate comprehensive information, enabling managers to base their actions on information instead of intuition. Users need fast, flexible access to current information, which they can look at from several perspectives.

Only on-line transaction processing systems and networks--people and programs simultaneously retrieving and updating information--satisfy the immediacy today's environment requires.

Response Statements	Numerical Assessment	
	1ST	2ND

1. The maintenance management system is computerized:
   - On-line (2 points)
   - Batch (1 point)
   - Not computerized (0 points) ( ) ( )

2. The computerized system integrates all maintenance informational needs:
   - Work Order Control (1 point) ( ) ( )
   - Work Planning (1 point) ( ) ( )
   - Work Measurement (1 point) ( ) ( )
   - Material Support and Control (1 point) ( ) ( )
   - Scheduling and Coordination (1 point) ( ) ( )
   - Preventive/Predictive Maintenance (1 point) ( ) ( )
   - Equipment History (1 point) ( ) ( )
   - Training Records and Scheduling (1 point) ( ) ( )
   - Budgetary Control (1 point) ( ) ( )
   - Cost Distribution (1 point) ( ) ( )
   - Management Reporting and Control (1 point) ( ) ( )
   - Supervisory/Administrative Support (1 point) ( ) ( )
   - Project Control (1 point) ( ) ( )

Appendix

– Downtime Reporting (1 point)	(	)	(	)
– Statistical Analysis (1 point)	(	)	(	)
– Trend Reporting (1 point)	(	)	(	)

3. The system satisfies real needs:
   – The right maintenance information is available to all levels
     of the organization, consistent with their needs. (0-2 points)   ( ) ( )
   – All personnel who need maintenance/material information
     are able to obtain it. (0-2 points)   ( ) ( )
   – Information is complete and reliable. (0-2 points)   ( ) ( )
   – Information and access to it, either on-line or in report format,
     is timely.(0-2 points)   ( ) ( )
   – System capabilities align well with user responsibilities.
     (0-2 points)   ( ) ( )
   – System security effectively controls who has access to specific
     systemfunctions and levels. (0-2 points)   ( ) ( )
   – Ad hoc report generation is available. (1 point)   ( ) (

4. Record files are complete and easy to locate. (0-2 points)   ( ) ( )

5. Monthly reports (i.e., budgetary) are promptly available relative
   to month-end:
   – One day or less (2 points)
   – Two to five workdays (1 point)
   – More than one week (0 points)   ( ) ( )

Element Summary:

Points Awarded	( )	( )
÷ Potential Points	35	35
= Current State of Effectiveness-Computer Support (Carry to 3 places)	0.	0.

Points Lost: 35 Potential Points – _____Points Awarded =_____

APPENDIX J

Assessment: Work/Job Planning

Sound planning is a prime requisite to effective control of the maintenance function. Maintenance planning is the advance preparation for selected jobs so they can be performed efficiently. It is the process of analyzing each job to determine such factors as what work is to be performed; the specific steps and sequence required; and for each step, the skills, skill level, labor-hours, parts, materials, drawings, special tools, equipment required with estimated costs.

Backlog measurement is a key planning tool to ascertain the degree to which maintenance keeps up with the amount of work requested and to adjust the size and make-up of the work force accordingly.

Response Statements	Numerical Assessment	
	1ST	2ND

1. The flow of work orders from originator to maintenance and within maintenance is clearly established and understood by all concerned parties. There are no piles of unprocessed work orders at any point in the system. This applies to emergency and standing work orders as well as backlog work orders. Work orders in the various stages of processing and execution are systemically filed (manual or computerized). (0-2 points)　　( ) ( )

2. The portion of each crew's capacity which is consumed by emergency response and routine duties, and which therefore is not available for backlog relief is known. (0-2 points)　　( ) ( )

3. Backlog data by priority code are effectively used for macro planning of resource requirements.
   ❑ At least once per period (3 points)
   ❑ At least once per quarter (2 points)
   ❑ On an ad hoc basis (1 point)
   ❑ Not used (0 points)　　( ) ( )

4. Planning predicts the workload of each skill group in sufficient detail to permit early identification of pending shortfalls, excesses, and imbalances; and in time to take appropriate corrective action. (0-2 points)　　( ) ( )

5. Backlog data is effectively utilized:
   ❑ Maintenance uses backlog to track how workload is changing. (0-2 points)　　( ) ( )
   ❑ The purpose of tracking work backlog is understood by all concerned parties. (0-2 points)　　( ) ( )

Appendix

❑ Through the use of work backlog, maintenance determines the needs for outside contractors.
  - During normal operations. (1 point)   ( ) ( )
  - During shutdowns. (1 point)   ( ) ( )
  - On major projects. (1 point)   ( ) ( )

6. "Ready Backlog" is distinguishable from backlog pending engineering,planning, material, shutdown, etc. (0-2 point   ( ) ( )

7. Backlog information is reliable and routinely utilized to make decisionsregarding:
  - Scheduling priority (1 point)   ( ) ( )
  - Internal resources vs. contractor (1 point)   ( ) ( )
  - Overtime (1 point)   ( ) ( )
  - Revise equipment operating plan (1 point)   ( ) ( )
  - Personnel reassignment or temporary help (1 point)   ( ) ( )

8. How many weeks of "Ready Backlog" currently exist?
  ❑ 3.0 to 5.0 weeks (4 points)
  ❑ 2.0 to 2.9 weeks (3 points)
  ❑ 5.1 to 7.0 weeks (2 points)
  ❑ Below 2.0 or above 7.0 weeks (1 point)
  ❑ Unknown (0 points)   ( ) ( )

9. How many weeks of "Total Backlog" currently exist, excluding annual turnabout?
  ❑ 6.0 to 8.0 weeks (3 points)
  ❑ 4.0 to 5.9 weeks (2 points)
  ❑ Below 4.0 or above 8.0 weeks (1 point)
  ❑ Unknown (0 points)   ( ) ( )

10. Policy specifies how much detailed planning of selected maintenance work will be done. A target has been established internally. (1 point) ( ) ( )

11. There is enough maintenance planning capacity that all work requiring planning can be planned. (0-2 points)   ( ) ( )

12. The concept of maintenance planning is fully accepted by maintenance andoperating supervision. (0-2 points)   ( ) ( )

13. Planning procedures are well documented.
  ❑ All normal repetitive procedures are documented (2 points)
  ❑ At least half of repetitive procedures are documented (1 point)
  ❑ No documentation (0 points)   ( ) ( )

14. Incumbent planners have been effectively trained in their responsibilities.
  - ❑ Combination of in-plant and outside training (2 points)
  - ❑ In-plant training only (1 point)
  - ❑ No training (0 points)   (   ) (   )

15. Current systems readily determine current status of all open work orders:
  - ❑ Completed work is identified. (1 point)   (   ) (   )
  - ❑ Scheduled work is identified. (1 point)   (   ) (   )
  - ❑ Ready Backlog is identified. (1 point)   (   ) (   )
  - ❑ Ready Backlog requiring downtime is identified. (1 point)   (   ) (   )
  - ❑ Waiting for shutdown is identified. (1 point)   (   ) (   )
  - ❑ Waiting for material is identified. (1 point)   (   ) (   )
  - ❑ Waiting for planning is identified. (1 point)   (   ) (   )
  - ❑ Waiting for engineering is identified. (1 point)   (   ) (   )
  - ❑ Waiting for approval is identified. (1 point   (   ) (   )

16. Backlog is well organized or easily retrievable (computerized systems).
  Work orders can be located by:
  - ❑ Assigned crew (1 point)   (   ) (   )
  - ❑ Priority (1 point)   (   ) (   )
  - ❑ Associated purchase order numbers (1 point)   (   ) (   )
  - ❑ Equipment (1 point)   (   ) (   )
  - ❑ Originator (1 point)   (   ) (   )

17. Work is requested with sufficient lead time (2 to 4 weeks) to allow
  effective planning:
  - ❑ Over 90% of the requests (5 points)
  - ❑ 80% to 90% of the requests (4 points)
  - ❑ 70% to 79% of the requests (3 points)
  - ❑ 60 to 69% of the requests (2 points)
  - ❑ 50 to 59% of the requests (1 point)
  - ❑ Less than 50% of the requests (0 points)   (   ) (   )

18. Coverage with planned work is high and is evident in the form of
  "Ready to Schedule" work orders listing work content by craft,
  materials, special tools andequipment, multi craft sequencing,
  required man-hours,crew size,and job site access:
  - ❑ Over 90% (5 points)
  - ❑ 75% to 90% (4 points)
  - ❑ 60% to 74% (3 points)
  - ❑ 40% to 59% (2 points)
  - ❑ 20% to 39% (1 point)
  - ❑ Less than 20% (0 points)   (   ) (   )

Appendix

19. All shutdown work is preplanned.
    - ❑ At least 90% of shutdown work (2 points)
    - ❑ Greater than 75% of shutdown work (1 point)
    - ❑ None (0 points)   (   ) (   )

20. To improve control, large jobs (over 40 hours) are broken into logical segments with separate work orders for each. (0-2 points)   (   ) (   )

21. Planners utilize effective methods, including:
    - ❑ Discussion with originator to establish objectives. (1 point)   (   ) (   )
    - ❑ Visiting job site to thoroughly determine job requirements. (1 point)   (   ) (   )
    - ❑ Determining safety/tag out/permit requirements. (1 point)   (   ) (   )
    - ❑ Discussion of job with maintenance supervision regarding necessityand correctness. (1 point)   (   ) (   )
    - ❑ Analysis and breakdown of job into steps and labor requirements. (1 point)   (   ) (   )
    - ❑ Determining all material requirements and preparing bills of material showing locations of reserved items. (1 point)   (   ) (   )
    - ❑ Obtaining or preparing necessary prints and sketches. (1 point)   (   ) (   )
    - ❑ Identifying and providing all special tools and equipment required tocomplete the job. (1 point)   (   ) (   )
    - ❑ Coordinating shop work with field requirements. (1 point)   (   ) (   )

22. A well-organized work order materials cage, under lock and key, contains necessary materials for all jobs which will be scheduled within the next fewweeks. All materials are tagged with the work order number.
    - ❑ Adequate security and control exists (1 point)   (   ) (   )
    - ❑ All materials are tagged with work order number (1 point)   (   ) (   )

23. Work measurement is part of the planning process. Accuracy is progressive.
    - ❑ Measurement is used regularly to define backlog, schedule jobs, measure efficiency and refine planning (2 points)
    - ❑ Measurement is applied but not effectively used for all purposes (1 point)
    - ❑ No measurement is being done (0 points)   (   ) (   )

24. Reference files are maintained to minimize planner time demands associated with repetitive jobs and effort:
    - ❑ Previous job plans. (1 point)   (   ) (   )
    - ❑ Material libraries. (1 point)   (   ) (   )
    - ❑ Labor libraries. (1 point)   (   ) (   )
    - ❑ Standard bills of material for repetitive jobs,overhauls, and rebuild. (1 point)   (   ) (   )

Appendix

25. Policy specifies the characteristics of jobs which must be
costed prior to authorization. (1 point)   (   ) (   )

26. Planners spend their day in essentially the following proportions:
- ❏ Site visitation (10-20%) (1 point)   (   ) (   )
- ❏ Job breakdown (10-15%) (1 point)   (   ) (   )
- ❏ Estimating (5-10%) (1 point)   (   ) (   )
- ❏ Material requirements (10-20%) (1 point)   (   ) (   )
- ❏ Material expediting (0-10%) (1 point)   (   ) (   )
- ❏ Inter/intra-coordination (10-20%) (1 point)   (   ) (   )
- ❏ Scheduling (10-20%) (1 point)   (   ) (   )
- ❏ Follow-up and reporting (5-15%) (1 point)   (   ) (   )
- ❏ Administration (0-10%) (1 point)   (   ) (   )

27. The use of formal planning has successfully reduced downtime
durations. (0-2 points)   (   ) (   )

28. Maintenance crews prefer to work on jobs which have been planned.
(0-2 points)   (   ) (   )

29. Maintenance supervisors substantiate planning effectiveness.
(0-2 points)   (   ) (   )
                                                              ——   ——

Element Summary:
    Points Awarded   (___) (___)
    ÷ Potential Points   100   100

    = Current State of Effectiveness-Work Planning   0.   0.
      (Carry to 3 places)

    Points Lost: 100 Potential Points - _____ Points Awarded =   _____   _____

APPENDIX K

ASSESSMENT: WORK MEASUREMENT

Work measurement is essential to the entire planning, scheduling and control process. Sophistication of work measurement technique varies with the maturity of the maintenance management installation.

Response Statements	Numerical Assessment 1ST	2ND

1.0 Work standards/estimates for maintenance work are established by:
- ❑ Engineered job standards. (5 points)
- ❑ Averages or estimates adjusted by work sampling. (4 points)
- ❑ Analytical estimates. (3 points)
- ❑ Historical averages. (2 points)
- ❑ Supervisor/planner estimates. (1 point)
- ❑ No measurement applied. (0 points)　　　　　　　　　( ) ( )

2. New jobs are estimated using slotting techniques in comparison to engineered benchmarks. (1 point)　　　　　　　　　( ) ( )

3. Estimates are automatically posted to repetitive jobs by the computer. (1 point)　　　　　　　　　( ) ( )

4. Planners can recover their logic and buildup of established standards. (1 point)　　　　　　　　　( ) ( )

5. Work measurement coverage is high:
- ❑ Over 85% of jobs (5 points)
- ❑ 70% to 84% (4 points)
- ❑ 50% to 69% (3 points)
- ❑ 30 to 49% (2 points)
- ❑ 1 to 29% (1 point)
- ❑ No work measurement (0 points)　　　　　　　　　( ) ( )

6. Standards/estimates are pre-applied but when feedback indicates a change of work scope, adjustments are post-applied. (1 point)　　　　　( ) ( )

7. Work measurement is a vital aspect of the managerial control process:
- ❑ The size and makeup of the maintenance work force has been determinedby careful measurement of the type and amount of work to be done--using work programs or a similar analytical format.(1 point)　　　　　　　　　( ) ( )

209

Appendix

❑ Management requests measurement of workload and required
work force before force adjustments (up or down) are made.
(1 point)   ( ) ( )
❑ Performance reports are regularly issued for the total department
and eachsupervisor. (1 point)   ( ) ( )
❑ Standards/estimates are effectively used to help control major
jobs. (1 point)   ( ) ( )
❑ The maintenance work force understands and accepts the
purpose of periodic manpower productivity measurement.
(1 point)   ( ) ( )

8. Records are maintained to facilitate modification of standards/
estimates as needed. Comparisons of actual to estimated hours are
regularly made and analyzed. One group of estimates, or estimates
by one planner may be inconsistent with others. (0-2 points) ( ) ( )

9. Methods and standards are integral to the maintenance management
program:
❑ There is sufficient confidence in and/or understanding of the
standards/estimates that they are posted to the crew copy
of the work order. (1 point)   ( ) ( )
❑ A full-scale methods effort is in place, with field review
of job activitiesby planners, engineers, and supervisors.
(1 point)   ( ) ( )
❑ Responsibility for researching and selecting best methods,
tools, materials,safety requirements and standards/estimates
is specifically assignedto one or more persons. (1 point)   ( ) ( )
❑ Proper tools and machine feeds and speeds are established
for MachineShop and Tool Room jobs. (1 point)   ( ) ( )
❑ Work sampling and other productivity studies focus attention on
the elimination or reduction of delays, non-productive time, and
productivity losses incurred for various reasons -- including
contractual agreement interpretation. (1 point)   ( ) ( )
❑ Shutdowns and turnarounds are studied to find means to
reduce their cost and duration, thereby improving on-stream time
and profitability. (1 point)   ( ) ( )

10. Maintenance is covered by an incentive program:
❑ Plant-wide Plan on Overall Operating Results (3 points)
❑ Small Group Plan (2 points)
❑ Individual Plan (1 point)
❑ No Plan (0 points)   ( ) ( )

210

Appendix

Element Summary:

Points Awarded
÷ Potential Points
$\underset{30}{()}$ $\underset{30}{()}$

= Current State of Effectiveness- Work Measurement   0.____ 0.____
  (Carry to 3 places)

Points Lost: 30 Potential Points- ____   Points Awarded   =   ____   ____

211

MATERIAL SUPPORT AND CONTROL

For maintenance to fulfill its mission, materials must be available when needed. Material availability and control of inventory can be conflicting objectives without managerial policy which best satisfies the overall needs of the plant, establishes a system for optimization, and calls for continual review and control.

Response Statements	Numerical Assessment	
	1ST	2ND

1. Purchasing, inventory and stores procedures covering stock, spare parts, special purchases, and "in-house" manufactured items are well documented by written instructions which have been effectively distributed to the involved organization(s). (1 point)   ( ) ( )

2. The supply room is sufficiently stocked to meet day-to-day needs as determined by the following service levels:
   - ❑ Insurance Spares (100%) - (1 point   ( ) ( )
   - ❑ Other Critical Spares (98%) - (1 point)   ( ) ( )
   - ❑ Standard Replacement Parts (95%) - (1 point)   ( ) ( )
   - ❑ Fasteners and Fittings (90%) - (1 point)   ( ) ( )
   - ❑ Small Tools (90%) - (1 point)   ( ) ( )
   - ❑ M&R Supplies (85%) - (1 point)   ( ) ( )

3. There is a perpetual inventory system in place with activity systematically recorded and reported to maintenance, purchasing and accounting for use in the management of inventory.
   - ❑ Computerized (2 points)
   - ❑ Manual (1 point)
   - ❑ No system (0 points)   ( ) ( )

4. Inventory control practices are effective:
   - ❑ Control increases with inventory value using A-B-C categorization.(1 point)   ( ) ( )
   - ❑ Cycle counting is used to preserve reliability of inventory records (1 point)   ( ) ( )
   - ❑ Free issue of low-value items, such as fasteners, is effectively Integrated into the system. (1 point)   ( ) ( )

5 An intelligent part numbering system is utilized. (1 point)   ( ) ( )

6. Usage records are employed to determine stocking levels, order points, and order quantities. (1 point)   ( ) ( )

Appendix

7. Maintenance, Purchasing, Accounting and the Supply Room work together to assure availability of necessary parts, elimination of obsolete parts, adjustment of stocking levels, minimal lost time by craftsmen, etc. (1 point)   (   ) (   )
   - ❑ Maintenance is regularly consulted regarding stocking parameters generated by the system. (1 point)   (   ) (   )
   - ❑ Human judgment can override the system on an item-by-item basis. (1 point)   (   ) (   )
   - ❑ Maintenance influences the type and quantity of repair materials to be set up in stock. (1 point)   (   ) (   )
   - ❑ A system exists which flags frequently ordered parts which must be authorized for stocking. (1 point)   (   ) (   )
   - ❑ Stocking levels (min/max), reorder points and order quantities are set up and maintained:
     – Automatically (2 points)
     – Manually (1 point)
     – Not set up (0 points)   (   ) (   )

8. There is an established procedure to review obsolescence, scrap, excessive quantities, etc. (2 points)   (   ) (   )
   - ❑ Zero and low activity items are reviewed at least once a year. (1 point)   (   ) (   )
   - ❑ Stores does not remove maintenance items from authorized stock without Maintenance agreement. (1 point)   (   ) (   )

9. Inventory categories are defined and segregated when analyzing inventory value and turns. (1 point)   (   ) (   )

10. Inventory turns by category are reasonable. (1 point)   (   ) (   )

11. Management supports the Scientific inventory Control (SIC) formula to optimize inventory management. Artificial decisions and targetsare not allowed to destroy the system. Order points and quantities are calculated on an economic basis. (1 point)   (   ) (   )

12. SIC formula parameters for EOQ calculation (carrying cost, purchase order cost, delivery time, etc.) are current and periodically updated. (1 point)   (   ) (   )

13. A computerized inventory control system automatically reorders when on-hand quantity drops to or below reorder point. (1 point)   (   ) (   )

14. There is an established insurance/critical spares program containing review provisions. They are protected from discard due to low usage. (1 point)   (   ) (   )

15. Withdrawal procedures are established and enforced. (0-2 points) ( ) ( )

16. An approved work order indicating which cost center to charge is sufficient authorization to withdraw stock materials. (1 point) ( ) ( )

17. Studies are periodically made of stock outages to keep tabs on levels and trends. (1 point) ( ) ( )

18. Maintenance has ready access to up-to-date material and stock information, including: part numbers, stock numbers, bin locations, economic order quantities, reorder points, etc.
   - ❑ Computerized (2 points)
   - ❑ Cataloged (1 point)
   - ❑ No Ready Access (0 points) ( ) ( )

19. Bill of Material information is cross-referenced by point-of-use. (1 pt) ( ) ( )

20. Attribute scanning on key words of the description are available to help find the desired items. (1 point) ( ) ( )

21. Procedures exist for reserving and kiting stock parts for planned jobs:
   - ❑ Reservation/Allocation (1 point) ( ) ( )
   - ❑ De-Allocation (1 point) ( ) ( )
   - ❑ Kiting or Pre-staging (1 point) ( ) ( )
   - ❑ Secured Staging to Avoid "Borrowing" from One Job for Another (1 point) ( ) ( )

22. Unit dollar values are readily available and routinely used in planning and work order approval efforts. (1 point) ( ) ( )

23. The cost of material consumed in the repair of specific units of equipment can be easily obtained. (1 point) ( ) ( )

24. Critical spare part lists exists for all important equipment. (1 point) ( ) ( )

25. There are control procedures in use for all company-owned tools and supplies, such as drills, special saws, ladders, files, gloves, etc.
   - ❑ Such items are identified. (1 point) ( ) ( )
   - ❑ Kept in designated places. (1 point) ( ) ( )
   - ❑ Checked in and out of secured tool crib to authorized personnel using a tool check or similar system. (1 point) ( ) ( )

26. Special tools are kept in good repair and in service. (1 point) ( ) ( )

27. There are standard listings of those tools which are company provided and those which are employee provided. (1 point) ( ) ( )

Appendix

28. Satellite inventories of shop-based materials, such as fasteners, O-rings lumber, and so forth:
    - ❏ Exist only by plan. There are no hidden satellite stocks throughout the plant. (1 point) ( ) ( )
    - ❏ Are under Stores control procedures. (1 point) ( ) ( )
    - ❏ Are located close to point-to-use. (1 point) ( ) ( )
    - ❏ Are limited to low value, fast-moving items. (1 point) ( ) ( )

29. Adequate material handling resources are available to stores to and from work sites. (1 point) ( ) ( )

30. There are also provisions for the mobilized delivery of urgent material needs, (i.e., policies relieve mechanics from the necessity to leave their job site to obtain parts or material while engaged in urgent repairs) therebyreducing material-related travel time by the crew involved in work performance. (1 point) ( ) ( )

31. Maintenance takes necessary lead time into consideration when requisitioning:
    - ❏ Stores Items (1 point) ( ) ( )
    - ❏ Purchased Items (1 point) ( ) ( )

32. Practices and procedures between Maintenance and Purchasing are clear and effective. (1 point) ( ) ( )
    - ❏ Maintenance is allowed to specify individual manufactures or suppliers when deemed necessary. (1 point) ( ) ( )
    - ❏ Maintenance is allowed to specify method of shipment when necessary. (1 point) ( ) ( )
    - ❏ Purchasing reviews any change in original specs with Maintenance and/or Engineering. They do not make material or equipment substitutions without consulting technically knowledgeable resources.(1 point ( ) ( )
    - ❏ Purchasing routinely, and when requested, follows up and expedites purchase orders. (1 point) ( ) ( )
    - ❏ There is an effective system for tracking Purchase Requisitions from generation to receipt by originator. (1 point) ( ) ( )
    - ❏ The volume of Emergency Purchase Orders is "reasonable". (1 pt) ( ) ( )
    - ❏ Production is not permitted to purchase direct and to maintain duplicate inventory of unit spares of special parts. (1 point) ( ) ( )
    - ❏ Maintenance is permitted to purchase direct locally when necessary. (1 point) ( ) ( )
    - ❏ When appropriate, Maintenance is permitted to contract work to local shops with Purchasing's cooperation. (1 point) ( ) ( )

Appendix

33. Blanket and system contracts, and blanket and system orders are effectively employed to minimize redundant paper work and administrative effort:
- ❑ Blanket Contracts (i.e. repetitive service) (1 point)   (   ) (   )
- ❑ System Contracts (i.e. service contractors negotiated by GO) (1 point)   (   ) (   )
- ❑ Blanket Orders (i.e. parts price agreement) (1 point)   (   ) (   )
- ❑ System Orders (i.e. parts costs negotiated by GO) (1 point)   (   ) (   )

34. Support systems for purchasing are effective:
- ❑ Quality performance measures have been established for materials. (1 point)   (   ) (   )
- ❑ Purchase Order tracking is functional. (1 point)   (   ) (   )
- ❑ A vendor rating system is in place and vendors are aware of their responsibility to provide materials of acceptable quality on time. (1 point)   (   ) (   )
- ❑ Procedures are in place which alert requisitioners regarding receipt of their materials. (1 point)   (   ) (   )
- ❑ Buyers are evaluated on the performance of vendors which they select. (1 point)   (   ) (   )
- ❑ There is an established interface/partnership with vendors to improve material quality, cost, availability and standardization at minimal carrying expense. (1 point)   (   ) (   )

35. There is an effective salvage, reclamation and unit repair program in place and operated as a profit center. (1 point)   (   ) (   )
- ❑ Lowest cost to the company governs the repair/replace decision. (1 point)   (   ) (   )
- ❑ Inventory controls are adequate & are under Stores control. (1 pt) (   ) (   )
- ❑ Storage is adequate. (1 point)   (   ) (   )
- ❑ Reclaimed items are returned to Stores' custody. (1 point)   (   ) (   )
- ❑ Stores accepts returned items in less-than-unit quantities for credit. (1 point)   (   ) (   )
- ❑ Work Orders for the salvage/repair of a replaced part is generated by the technician who replaced it. (1 point)   (   ) (   )

36. There is appropriate storage provision for left over materials (short lengths, etc.), scrap and waste. (1 point)   (   ) (   )

37. A two-bin system is utilized by Stores to facilitate issue while maintaining control on high-volume, low-cost items. (1 point)   (   ) (   )

38. Storeroom layout and procedures maximize order fulfillment service and functional efficiency:
- ❑ Parts are identified with stock (computer ID) numbers. (2 points) (   ) (   )
- ❑ Stock storage is environmental controlled as needed. (1 point)   (   ) (   )

216

Appendix

❑ If needed, rotating equipment in stock is regularly rotated to avoid development of bearing flat spots. This is controlled by a PM-type revolving file. (1 point)   (   ) (   )

❑ ABC analysis has been applied to the arrangement of stock storage to optimize efficiency of stock retrieval. (1 point)   (   ) (   )

❑ Stock pick lists are arranged to facilitate picking. (1 point)   (   ) (   )

❑ Computer control will accept no more than two storage locations for any one item in any given storeroom. The impact of this is to keep inventory from becoming too spread out and impossible to control. (Note: MP2 allows more than two storage locations; therefore award no points) (1 point)   (   ) (   )

❑ QA inspection is required, in varying degrees based on criticality, for all materials received for stock or special orders. (1 point)   (   ) (   )

❑ Bar coding is utilized to facilitate tracking of Purchase Orders, Stock Requisitions, parts, and progress through the receipt/ inspection/stocking/issue process. (1 point)   (   ) (   )

39. In Central Stores, parts are stored randomly to maximize storage density.(1 point)   (   ) (   )

❑ In satellite stores, parts are stored by equipment to facilitate retrieval. (1 point)   (   ) (   )

❑ Cannibalization from a serviceable unit (generally not installed) to another serviceable unit (generally installed) is against established policy. When necessary in an emergency situation, the necessary work orders, Stores requisitions, and/or purchase orders are initiated to restore the cannibalized unit to a serviceable state. (1 point)   (   ) (   )

Element Summary:
Points Awarded   (___) (___)
÷ Potential Points   100   100

Current State of Effectiveness- Material Support   0.   0.
    (Carry to 3 places)

Points Lost: 100 Potential Points - ___Points Awarded = _____   _____

217

APPENDIX M

ASSESSMENT: SCHEDULING AND COORDINATION

Scheduling and associated coordination is the process by which all required resources are made available at the proper time and place, with necessary access, so that work can be started and processed to completion with minimal delay and within established time frames. The established time frames reflect available manpower and the relative importance of work to be performed. Resources encompass manpower, materials, tools, equipment, and reference materials. Access refers to equipment in accessible and safe (locked out) state, including any necessary permits.

Response Statements	Numerical Assessment	
	1ST	2ND
1. The criticality of effective scheduling is recognized and supported throughout the organization. (0-2 points)	( )	( )
2. Scheduling procedures and timing are well documented and distributed. (0-2 points)	( )	( )
3. Work activity is analyzed by priority code to determine trends of man-hours spent on scheduled work. (0-2 points)	( )	( )
4. The relationship of backlogged work, and backlog relief experience is regularly analyzed and reviewed with management. ❑ At least monthly (2 points) ❑ At least quarterly (1 point) ❑ Never or rarely reviewed (0 points)	( )	( )
5. No more than 20% of the backlogged work orders on the MP2 Work Order Aging Report are over 60 days old. (1 point)	( )	( )
6. The system of work prioritization effectively distinguishes between legitimate rush jobs, and those which can be planned.		
❑ The system for prioritizing work orders has been thoroughly developed in consultation with operating personnel. (0-2 points)	( )	( )
❑ The system reflects equipment importance as well as individual job importance. (0-2 points)	( )	( )
❑ All work orders are assigned a priority code. (0-2 points)	( )	( )
❑ Work order age is also considered in the process of selecting backlog to be scheduled. (0-2 points)	( )	( )
❑ Operating/Maintenance liaisons are individualized and effectively utilized to establish finite priorities prior to each scheduling period. (0-2 pts)	( )	( )

218

Appendix

- ❑ There is seldom any confusion about the relative priorities of what work Maintenance should be doing. (0-2 points)   ( ) ( )
- ❑ Originators are not allowed to "cry wolf" to get their work performed more quickly. The Plant Manager supports this policy and the Maintenance manager polices it. (0-2 points)   ( ) ( )
- ❑ The Plant Manager and other senior members of management do not abuse the privileges of rank to override the system. (1 point)   ( ) ( )

7. Scheduled job start, completion and crew assignments are publishedand distributed. (0-2 points)   ( ) ( )

8. Schedule performance is measured and reported. Compliance and percentof man-hours scheduled are calculated, analyzed, plotted, published, and regularly reviewed by management. (1 point)   ( ) ( )
   - ❑ Both graphs and data are shared or posted within the department and with production plus other customers. (3 point)
   - ❑ Either graphs or data are shared or posted within the department and with production plus other customers. (2 point)
   - ❑ Shared or posted only within the department. (1 point)
   - ❑ Not posted or shared; only required KPI report is sent to GO.(0 pts)( ) ( )

9. Percent of labor resource scheduled is high, include PLANNED and PPM work only:
   - ❑ 90% to 100% (10 points)
   - ❑ 85% to 89% (9 points)
   - ❑ 80% to 84% (8 points)
   - ❑ 75% to 79% (7 points)
   - ❑ 70% to 74% (6 points)
   - ❑ 60% to 69% (5 points)
   - ❑ 50% to 59% (4 points)
   - ❑ 40% to 49% (3 points)
   - ❑ 30% to 39% (2 points)
   - ❑ 1% to 29% (1 point)
   - ❑ No measurable scheduling (0 points)   ( ) ( )

10.Schedule compliance, substantiated by the KPI report, is high:
   - ❑ 90% to 100% (10 points)
   - ❑ 85% to 89% (9 points)
   - ❑ 80% to 84% (8 points)
   - ❑ 75% to 79% (7 points)
   - ❑ 70% to 74% (6 points)
   - ❑ 60% to 69% (5 points)
   - ❑ 50% to 59% (4 points)
   - ❑ 40% to 49% (3 points)

□ 30% to 39% (2 points)
   □   1% to 29% (1 point)
   □   No measurable scheduling (0 points)   (   )(   )

11. Procedures for requesting, coordinating and controlling supporting services, such as mobile equipment, rigging, transportation, etc., are effective. (0-2 pts)   (   )(   )

12. Maintenance schedules are closely coordinated with Production schedules, and availability dates for repair of Production equipment are agreed upon in advance.
   □   Initial and final coordination each week. (3 points)
   □   Weekly (2 points)
   □   Occasionally (1 point)
   □   No coordination or agreement on schedules (0 points)   (   )(   )

13. There are effective procedures for assigning work to specific mechanics. (0-2 points)   (   )(   )
   □   Where a work assignment board is appropriate, the supervisor maintainsenough work in each mechanic's slot to keep him gainfully assigned. (0-2 points)   (   )(   )

14. Those procedures include consistent effort to assign mechanics to work which matches their skill classification and which is supportive of the OJT needs. (0-2 points)   (   )(   )

15. All types of backlog work (repair, changes, and additions) are scheduled but only those jobs with all material on hand appear on the current schedule. (0-2 points)   (   )(   )

16. The weekly schedule is reviewed, finalized and approved during a joint maintenance/production meeting held late in the week so that both weekendwork and next week's work can be addressed. Production liaisons, Maintenance Supervisors, and Maintenance Planners attend regularly. Others attend as warranted. Schedules are written, approved by the Plant Manager or his designee and then distributed. (0-2 points)   (   )(   )

17. Production and Maintenance diligently work together toward high schedule compliance:
   □   Production and Maintenance reach agreement on the weekly schedulebefore work is performed. (0-2 points)   (   )(   )
   □   Production consistently makes equipment available according to the approved schedule. (0-2 points)   (   )(   )

❑ Maintenance continues to coordinate throughout the schedule week to assure that all agreements and arrangements are still valid. Nothing is assumed even though the schedule has been approved. (0-2 points) ( ) ( )

❑ Maintenance Supervisors make every effort to follow schedule every day. (0-2 points) ( ) ( )

❑ Jobs-in-progress are reviewed against the schedule every day. (0-2 points) ( ) ( )

❑ Causes of deviation from schedules and deviation from estimates aredetermined and addressed. (0-2 points) ( ) ( )

❑ Production personnel do not assign work to Maintenance craftsmen. Even craftsmen assigned to certain production are as receive regular instructions from their Maintenance Supervisor, and respond to production requests only in the event a machine goes down. (0-2 points) ( ) ( )

❑ Analysis is made of work orders to determine the degree to which priority is abused. (0-2 points) ( ) ( )

❑ Analysis is made of completed work orders to determine the degree to which the priority implied completion date is met. (1 point) ( ) ( )

18. Production, production scheduling, and maintenance all contribute to adherence to the PM schedule.

❑ Production notifies Maintenance, with as much advance notice as possible, of any maintenance or other downtime (over a few minutes) which occurs due to:-Changes in the production schedule. (0-2 pts) ( ) ( )

❑ Unexpected equipment shutdowns. (0-2 points) ( ) ( )

❑ Production planning is totally involved by scheduling production equipment downtime for PM. They take advantage of product changeovers and other necessary outages to help keep the PM program on schedule with the least demand on near-term production capacity. (0-2 points) ( ) ( )

❑ Maintenance, with notification by Production, and through use of backlog by machine, takes advantage of downtime to do any possible jobs, including backlog and PM inspections. (0-2 points) ( ) ( )

❑ Schedule provision is made to take advantage of those "windows". Schedule compliance does not suffer. Plan "B" is accomplished, rather than Plan "A". (0-2 points) ( ) ( )

19. The Maintenance organization recognizes and accepts accountability (partial) for meeting Production schedules. (0-2 points) ( ) ( )

20. Maintenance management regularly attends and contributes to plant operating and planning meetings. (0-2 points) ( ) ( )

21. Peak loads and resource "valleys" are balanced:
- ❑ Shutdowns and/or other major projects are planned sufficiently in advance to permit effective allocation or procurement of manpower, materials and tools. (0-2 points)   ( ) ( )
- ❑ Concurrent peak loads (overhauls and construction) are avoided if at all possible. (1 point)   ( ) ( )
- ❑ Vacation schedules are determined far ahead, and in keeping with maintenance planning of known peak loads. (1 point)   ( ) ( )

Element Summary:

Points Awarded   (___) (___)
÷ Potential Points   100   100

Current State of Effectiveness-Coordination and Scheduling   0.___ 0.___
(Carry to 3 places)

Points Lost: 100 Potential Points -_____ Points Awarded =   _____ _____

APPENDIX N

ASSESSMENT SUMMARY & COMPARISON TO BENCHMARKS

	Organ.	Comp. Supp.	Planning	Measurement	Material Supp.	Scheduling	Composite
Potential Pts.	60	35	100	30	100	100	425
Pts. Awarded							
Score *							
Data Base Low	0.180	0.000	0.006	0.000	0.130	0.090	0.078
Data Base Avg	0.625	0.500	0.382	0.288	0.584	0.418	0.475
Data Base High	0.950	0.929	0.850	0.717	0.890	0.880	0.877
Proactive Target	0.700	0.750	0.700	0.600	0.700	0.800	0.720
World Class Goal	0.850	0.850	0.800	0.750	0.800	0.850	0.820

*Score = Points Awarded/Potential Points

Glossary

Maintenance preparatory terms (planning, estimating, procurement, coordination and scheduling).

Autonomous Maintenance: Routine maintenance and PM's are carried out by operators in independent groups. These groups , which may include maintenance workers solve problems without management intervention. The maintenance department is only officially called for bigger problems that require more resources, technology or downtime.

Asset: A machine, building or system.

Asset number: A unique number necessary to identify an asset.

Backlog: All plannable work not yet complete.

Call Back Job: Job to which maintenance must return because the asset failed again for the same reason because the job was not performed properly the first time.

Charge rate: This is the rate in dollars charged for a mechanic's time. Rates used may be base direct wages or fully loaded to include benefits and overhead (such as supervision, clerical support, shop tools, truck expenses, supplies). Rates used may be average for a maintenance group or specific for a given mechanic.

Corrective maintenance: Restorative maintenance activity initiated as a result of finding during a scheduled inspection.

Deferred maintenance: This is known work that should be performed but is put off indefinitely, usually due to budgetary constraints.

DIN crew: Do It Now. A portion of the maintenance crew designated to perform urgent work.

Glossary

Emergency work: Maintenance work requiring immediate response from the maintenance staff. Usually associated with some kind of danger, safety, damage or failure of critical production equipment.

Failure Code: Why did the part fail (broken, worn through, bent, etc.).

Feedback: Information provided by assigned mechanic or supervisor that will enhance future planned job packages.

Inspectors: Those technicians with primary responsibility for PM/PdM inspections.

Intrusive task (also interruptive task): Any inspection that interrupts normal operation of an asset.

MTBF: Mean time between failures. Metric used to evaluate reliability of an asset.

MTTR: Mean time to repair. Metric used to evaluate the effectiveness of job preparation. A trend line is necessary to convey improvement.

Maintenance: The dictionary definition is "the act of holding or keeping in a preserved state."

Maintenance Prevention: Maintenance-free designs resulting from increased effectiveness in initial design of equipment.

PCR: Planned Component Replacement based upon known life of component.

PM: Preventive Maintenance. A series of tasks to extend the life of an asset or to detect that an asset has reached a point of critical wear and is about to fail.

PM Clock: The parameter that initiates a PM routine to be scheduled.

PM frequency: Time interval driving PM inspections.

PdM Predictive Maintenance techniques that inspect asset condition relative to a known point of failure, thereby predicting when failure will occur.

Priority: The relative urgency of a job.

Proactive: Action before a stimulus. Preparation for effective execution prior to failure.

Glossary

Root cause (and root cause analysis): The underlying cause of a problem.

Routine work: Work of a known content at a known frequency.

Task: Single item in a step-by-step procedure.

Technical Library: Repository of all maintenance information filed by asset; including maintenance manuals, drawings, repair history, vendor catalogs, shop manuals, etc.

Work Order: Written authorization to proceed with a repair or other activity to be performed by the maintenance organization.

Work request: Formal request to have work performed. Subsequently transformed into a Work Order upon authorization.

Index

Index